Data and Argumentation in Historical Pragmatics

Pragmatic Interfaces

Series Editors:
Enikő Németh T., University of Szeged
Dániel Z. Kádár, Hungarian Academy of Sciences
Károly Bibok, University of Szeged

In the last two decades it has become increasingly clear that language and language use cannot be studied separately and independently of each other. This new approach assumes an interaction between grammar (phonology, morphology, lexicon, syntax and semantics) and pragmatics. An analysis of the interfaces between each component of grammar and pragmatics (the 'interface view') can also be applied to hard-pragmatics and soft-pragmatics research. Hard-pragmatics studies the field of language use from philosophical, linguistic and logical points of view, while soft-pragmatics explores phenomena of language use from a social and socio-cultural perspective.

The definitions hard- and soft-pragmatics, adopted around the 1980s, have become somewhat dated since pragmatics has become a field of its own, and so these two trends have merged to some extent. Also, various pragmaticians made important attempts to blend these approaches. Nevertheless, a border between these areas continues to exist: hard-pragmaticians rarely venture into socio-pragmatic issues, and, vice versa, soft-pragmatic studies rarely make use of formal tools of hard-pragmatics.

Pragmatic Interfaces fills an important knowledge gap in the field of pragmatics as the first major publication project devoted to studying grammar–pragmatics interfaces and merging of soft-pragmatics with hard-pragmatics. Through this merging many pragmatic phenomena could be essentially revisited. *Pragmatic Interfaces* follows an interdisciplinary approach, allowing scholars from different areas of grammar and pragmatics to collaborate.

Published:
Face and Face Practices in Chinese Talk-in-Interaction: A Study in Interactional Pragmatics
Wei-Lin Melody Chang
Implicit Subject and Direct Object Arguments in Hungarian Language Use: Grammar and Pragmatics Interacting
Enikő Németh T.
Impoliteness in Corpora: A Comparative Analysis of British English and Spoken Turkish
Hatice Celebi
Politeness Phenomena across Chinese Genres
Edited by Xinren Chen

Data and Argumentation in Historical Pragmatics: Grammaticalization of a Catalan Motion Verb Construction

Katalin Nagy C.

SHEFFIELD UK   BRISTOL CT

Published by Equinox Publishing Ltd.

UK: Office 415, The Workstation, 15 Paternoster Row, Sheffield, South Yorkshire S1 2BX
USA: ISD, 70 Enterprise Drive, Bristol, CT 06010
www.equinoxpub.com

First published 2019
© Katalin Nagy C. 2019

British Library Cataloguing-in-Publication Data
A catalogue record for this book is available from the British Library.
ISBN-13 978 1 78179 746 4 (hardback)
ISBN-13 978 1 78179 747 1 (ePDF)

Library of Congress Cataloging-in-Publication Data
Names: Nagy C., Katalin, author.
Title: Data and argumentation in historical pragmatics : grammaticalization
   of a Catalan motion verb construction / Katalin Nagy C.
Description: Sheffield, UK ; Bristol, CT : Equinox Publishing Ltd, 2019. |
   Series: Pragmatic interfaces | Includes bibliographical references and
   index.
Identifiers: LCCN 2018047848 (print) | LCCN 2019000864 (ebook) | ISBN
   9781781797471 (ePDF) | ISBN 9781781797464 (hb)
Subjects: LCSH: Catalan language--Verb. | Motion--Terminology.
Classification: LCC PC3897.V2 (ebook) | LCC PC3897.V2 N35 2019 (print) | DDC
   449/.956--dc23
LC record available at https://lccn.loc.gov/2018047848

Typeset by Steve Barganski

Contents

Acknowledgments

This volume is a revised version of my PhD thesis, submitted to the University of Szeged (Hungary), Doctoral School in Linguistics in 2013. It would not have been possible to write this book without the support and inspiration I have received from several people. It is a pleasure and honour to acknowledge their contributions. My interest on the topic of this volume began during my studies at the Departments of Hispanic Studies and of General Linguistics at the University of Szeged, where I had wonderful teachers of linguistics, especially of pragmatics, comparative Romance linguistics, Catalan and Spanish linguistics and historical linguistics. I am very grateful to my university teachers.

First of all, I would like to thank my PhD supervisor, Enikő Németh T., whose patient guidance greatly enhanced the quality of this book. I am grateful to her for her friendship and continuous support for my PhD studies and related research, constructive discussions of several topics which are touched upon in this volume, and for encouraging me to transform the thesis into this book. In addition, I am grateful to Tibor Berta for his continuous help. He generously put at my disposal some medieval Catalan and Spanish texts. I wish to thank the opponents of my thesis, Csilla Ilona Dér and Ildikó Szijj, for their valuable comments and critiques on a previous draft. I would like to say a warm thank you to my first Spanish teacher Éva Kertai and to commemorate, with love, Eloi Castelló Gassol, who taught me my first words in Catalan.

Work on the present book was supported by the Research Group for Theoretical Linguistics of the Hungarian Academy of Sciences at the Universities of Debrecen and Szeged. I would like to express my sincere gratitude to all my colleagues in the Research Group, and special thanks must go to the director of the Research Group, András Kertész, and Csilla Rákosi for their personal and academic support. Their many helpful comments on previous manuscripts had a strong influence on the organization and content of this volume. I am grateful to Anna Fenyvesi, George Seel, and Zsuzsanna Németh for patiently improving my English. All the remaining errors are my own. I would also like to thank Tibor Szécsényi for all his technical help.

Some parts of this book are based on articles and book chapters published earlier. Thanks to Peter Lang GmbH for the kind permission to republish materials from the following paper (Copyright Peter Lang International Academic Publishers, permission conveyed through Copyright Clearance Center, Inc.):

Nagy C., Katalin (2008) Data in historical pragmatics: A case study on the Catalan periphrastic perfective past. In András Kertész and Csilla Rákosi (eds.) *New Approaches to Linguistic Evidence: Pilot Studies* (MetaLinguistica 22) 171–197. Frankfurt am Main: Lang.

I would also like to thank de Gruyter Mouton for the kind permission to reuse the following text (Copyright de Gruyter Mouton, permission conveyed through Copyright Clearance Center, Inc.):

Nagy C., Katalin (2010) The cognitive background of grammaticalization. In Enikő Németh T. and Károly Bibok (eds.) *The Role of Data at the Semantics-Pragmatics Interface* (Mouton Series in Pragmatics 90) 207–260. Berlin: Mouton de Gruyter.

And, finally, I would like to express my gratitude to John Benjamins Publishing Company for the kind permission to reprint the following papers (Copyright John Benjamins Publishing Company https://www.benjamins.com/catalog/jhp; https://benjamins.com/catalog/slcs.153):

Nagy C., Katalin (2010) The pragmatics of grammaticalisation: The role of implicatures in semantic change. *Journal of Historical Pragmatics* 11(1): 67–95.

Nagy C., Katalin (2014) Methods and argumentation in historical linguistics: A case study. In András Kertész and Csilla Rákosi (eds.) *The Evidential Basis of Linguistic Argumentation* (Studies in Language Companion Series 153) 71–102. Amsterdam: John Benjamins.

I would like to express my gratitude to the series editors Enikő Németh T., Dániel Z. Kádár, and Károly Bibok for their work with the manuscript. Further acknowledgements are due to Janet Joyce, Valerie Hall, and Steve Barganski for their assistance during the publication process.

Last but not least, a warm thanks to my family: my parents, my sister, my husband Imre, and our daughters Linda and Boglárka for their love, encouragement, and tolerance throughout the writing of this book.

Abbreviations

1	first person
2	second person
3	third person
Art	article
Asp	aspectual
Aux	auxiliary
Clit	clitic
Ger	gerund
ImpP	imperfect past tense
Inf	infinitive
Neg	negative particle
Part	participle
PerfP	perfect past tense (preterit)
Pl	plural
Prep	preposition
Pres	present tense
Refl	reflexive pronoun
Sg	singular
Sub	subjunctive
Temp	temporal

CHAPTER 1

Introduction

1.1 Motive of the research

In his pioneering paper on the feasibility of historical pragmatics, Jucker (1994) argued convincingly for an approximation of pragmatics and historical linguistics (see also Jacobs and Jucker 1995). Since then, there have been many advances in historical pragmatics to explore the pragmatic motivations of language change, including research on semantic change in grammaticalization (cf. Hopper and Traugott 1993; Traugott 1999; Kearns 2002; Fortson 2003; Traugott and Dasher 2002/2004; Bybee 2002 and 2003/2005; Narrog and Heine 2011; Dér 2008 and 2013, among others). In historical pragmatics, grammaticalization is regarded as a discursive-pragmatic phenomenon, thus going beyond the traditional concept of grammaticalization that concentrates on grammatical aspects of the process[1] (cf. Hopper and Traugott 1993: 2; Alonso 1982: 194; Pons Bordería and Ruiz Gurillo 2001: 317). According to Heine, Claudi, and Hünnemeyer (1991), two main lines of research can be identified within grammaticalization studies, characterizable as macro- vs. micro-levels. The micro-level approach highlights the importance of examining the concrete contexts in which grammaticalization takes place and explores pragmatic inferences that have an influence on semantic change. These studies emphasize that the regularity of semantic change cannot be accounted for without paying attention to the implicatures and inferences that arise in language use and later become attached to a certain construction by a process of conventionalization in the course of language change (Hopper and Traugott 1993; Traugott 1999; Levinson 2000). In contrast, the macro-level approach explores cross-linguistically recurrent pathways of semantic change, i.e. the 'cognitive paths' along which meanings evolve in the different languages of the world (Bybee 2002; Bybee, Perkins, and Pagliuca 1994; Heine and Kuteva 2002). From this perspective, the long-term process of semantic change is seen as originating in various cognitive mechanisms.

This book discusses a particular process of semantic change: that of the Catalan *anar* 'go' + infinitive construction. Catalan is special among Romance languages in that a purposive construction involving this motion verb achieves a perfective past meaning. This evolution is in contrast with the cross-linguistically recurrent pathway of verbs meaning 'go' to grammaticalize into future markers in similar constructions, detected in several languages of the world (cf. Badia i Margarit 1981; Nadal and Prats 1982/1996; Pérez Saldanya 1996: 76–77; 1998). Although this volume discusses a micro-level study on a grammaticalization process which does not fit any cross-linguistically frequent tendency hitherto revealed, it will hopefully also provide new information about the macro-level process of semantic change of motion verbs in general.

Historical pragmatics research – as well as synchronic pragmatics – concerns levels of meaning which are difficult to access (e.g. implicatures, inferential contents), and thus we cannot avoid the problem of data use and methodology. Issues concerning data, evidence, methods, and argumentation in linguistic studies have been central topics over the last few decades, not only in contemporary linguistics in general (see Lehmann 2004; Kepser and Reis 2005; Penke and Rosenbach 2004/2007; Kertész and Rákosi 2008, 2012, 2014a, 2014b), but also in historical linguistics and pragmatics (see Fischer 2004, 2007; Lass 1997; Jucker 2009), and, moreover, in historical pragmatics (see Jucker 1994; Jacobs and Jucker 1995; Fitzmaurice and Taavitsainen 2007; Navarro 2008). Methodological issues have not only been present in historical pragmatics since the beginnings (cf. Jucker 1994; Jacobs and Jucker 1995), but their resolution is seen as a crucial factor which will determine the future of this linguistic sub-discipline (Fitzmaurice and Taavitsainen 2007: 29). The present book aims to examine some methodological questions in historical pragmatics and contribute in this way to recent metalinguistic discussions on data, methods, and argumentation in linguistics.

1.2 Aims of the research

The present research has two interrelated aims, one empirical and one theoretical. Its object-scientific aim is to examine an actual historical linguistic phenomenon: the beginnings of a semantic change process in the grammaticalization of the medieval Catalan *anar* + infinitive construction (*infinitive* hereafter abbreviated as *Inf*). Its theoretical aim is to examine certain questions concerning historical pragmatics methodology by looking at this research process from the outside. My research questions in general are the following (specific object-scientific and methodological research questions will be presented in Sections 2.4.3 and 3.1 respectively):

Object-scientific research questions:

1. What patterns of different morphological variations of the construction 'go' + Inf can be detected in the historical corpus, and which variant should we take as the starting point in the semantic change process?
2. What meaning might the Catalan construction *anar* + Inf have conveyed at an intermediate stage of its grammaticalization?
3. How and in what contexts might the semantic change of the construction have taken place?

Methodological research questions:

1. What data and methods does historical pragmatics use in the study of semantic change in grammaticalization?
2. What role does context play in semantic change in grammaticalization and in historical pragmatics methodology?
3. Can metatheoretical considerations be useful when deciding between rival hypotheses during research?

1.3 Methodology of the research

1.3.1 A comparative approach

The study presented in this book has been carried out relying on data and methods traditionally accepted in historical pragmatics, and especially on a historical corpus of Catalan and Spanish texts from the 13th to 16th centuries. Occurrences have been analysed qualitatively in their broad contexts, although I have applied quantitative methods as well (cf. Navarro 2008). I have employed a comparative approach in order to collect as much relevant data as possible.

First, I have compared the Catalan construction under study with a parallel medieval Spanish construction. Similar motion verb constructions were used in the Middle Ages, not only in Catalan, but in various Western Romance languages (cf. Colon 1976/1978). A comparison with Spanish seemed to be useful, because the analogous Spanish GO-construction (*ir* 'go' (*a*) + Inf)[2] grammaticalized into an immediate future tense, that is, in contrast to the Catalan construction, its evolution fits a cross-linguistically frequent pathway of semantic change in motion verb constructions. The development of future-tense makers from GO-verbs in purposive constructions is a well-documented grammaticalization process, revealed in various languages around the world. The concept of 'motion' is mentioned in Bybee and Pagliuca (1987), and also in Bybee et al. (1994) as a typical source of a future tense (for further examples, see Heine and Kuteva 2002: 161–

163). Given the close relationship between the Catalan and Spanish languages and the great degree of similarity of the GO-constructions in the medieval stages of both languages (see Section 2.2), I have assumed that a comparison of these medieval constructions can provide useful data about the untypical development of the Catalan *anar* + Inf construction.

Second, relying on results of relevant literature, I have compared the Catalan GO-construction with further medieval periphrases used in similar contexts. A possible path for motion verbs to develop into future-tense morphemes is provided in Bybee (2002: 181) as follows: *motion with the aim of doing something > intention > future* (see also Bybee et al. 1994: 266–270). At an intermediate stage of this pathway the construction comes to be used to express the subject's intention to do something (cf. Bybee et al. 1994: 254; Heine and Kuteva 2002: 161–163). In order to examine whether the development of the Catalan construction followed a similar semantic path, I have included in the research occurrences of intentional periphrases (namely, the medieval Catalan *pensar (de)* + Inf and the medieval Spanish *pensar (de)* + Inf[3] constructions). I also have compared the use of Catalan *anar* + Inf with that of the inchoative construction *començar (a/de)* 'begin to' + Inf,[4] used in similar contexts in the language stage under study. This comparison has been motivated by the fact that some authors connect the use of the medieval *anar* + Inf construction with an inchoative meaning (see the inchoative hypothesis in Section 2.3.3).

In conclusion: the research covers not only occurrences of the Catalan *anar* 'go' + Inf, whose semantic development is our main focus in this book, but also those of the following periphrases: the medieval Catalan *començar (a/de)* 'begin to' + Inf and *pensar (de)* 'think of' + Inf, and the medieval Spanish *ir (a)* 'go' + Inf, *començar (a/de)* 'begin to' + Inf and *pensar (de)* 'think of' + Inf.

As for the methodological aim of this book, I rely on Kertész and Rákosi's (2008 and 2012) metalinguistic model, the so-called *p-model* (presented in Section 3.6.2).

1.3.2 Historical corpus of the study

The study presented here has been carried out relying on data and methods traditionally accepted in historical pragmatics, and especially on a historical corpus of Catalan and Spanish texts from the 13th to 16th centuries. As mentioned above, I have searched the corpus for occurrences of the following constructions: Catalan *anar* 'go' + Inf, *pensar (de)* 'think of' + Inf and *començar (a/de)* 'begin to' + Inf, and Spanish *ir* 'go' *(a)* + Inf, *pensar (de)* 'think of' + Inf and *començar (a/de)* 'begin to' + Inf. In addition, I have used further data sources accepted in historical pragmatics, which I present in the methodological chapter.

The hypotheses provided in the previous literature on the evolution of the Catalan *anar* + Inf construction are based primarily on the analysis of occurrences found in three famous Catalan medieval chronicles. Bruguera (1981) and Juge

(2002, 2006, 2008) work with examples taken from the Chronicle of James I. Desclot's Chronicle is the data source for Mendeloff (1968), Pérez Saldanya (1996), Pérez Saldanya and Hualde (2003), and some parts of it for Juge (2002, 2006, 2008), while Muntaner's Chronicle is the source for Mendeloff (1968). Berchem (1968) bases his analysis on examples taken from *L'Atlas italo-suisse*, while Detges (2004) did not make reference to any historical corpus. He takes his examples from previous literature, including occurrences originating in Desclot's and Muntaner's chronicles.

The broadest range of occurrences is analysed by Colon (1959/1978 and 1976/1978), including many occurrences taken from other Romance languages. The construction *anar* + Inf is missing in more than half of the 46 texts examined, but even so, he can find almost 500 occurrences. A similarly broad corpus is taken into consideration in Segura-Llopes (2012), who analyses occurrences of the GO-construction for a grammar of Old Catalan, as part of a project to compile a historical corpus. In summary, the analyses presented in previous literature, except for Colon (1959/1978 and 1976/1978) and Segura-Llopes (2012), work with a more limited range of occurrences than the research presented in this volume. One of the novelties of the present analysis is that it takes together occurrences found in the three famous medieval Catalan chronicles and analyses them using a pragmatic approach. I take into account 427 occurrences of the medieval Catalan construction *anar* + Inf, which exceeds the amount of examples examined in the previous literature. My corpus covers a longer period, thus it provides a basis for a more thorough comparison than studies based on one or two historical texts.

1.3.2.1 Catalan texts in the corpus

Throughout this volume, the Catalan texts[5] of the corpus are referred to by abbreviations assigned to them as follows:

(Jau)   *Llibre dels fets del rei en Jaume*   'The Book of the Deeds of James I of Aragon'

The autobiography of James I is the oldest of the famous chronicles written in the Catalan language. Since the writing of the book was inspired by the capture of Majorca (1229), it can be dated between 1229 and 1276, the date of James's death. However, the surviving manuscripts can be dated to a later period, the oldest one being from 1343. Although this date is posterior to the supposed time of writing of the other two chronicles, this text reflects the most archaic language state (cf. Bruguera 1981). Although James I was learned, he probably could not write, and the book was written by scribes at his dictation. James expected the work to be read out loud. The language of the chronicle is colloquial, with features of oral style, written in the first person.

(Desc)   Desclot, Bernat: *Crònica*   'Chronicle'

The second great medieval Catalan chronicle is that of the royal chancellor Bernat Desclot, officially titled *Llibre del rey en Pere de Aragó e dels seus antecessors passats* ('Book of King Peter of Aragon and of his Ancestors'). It deals with the reign of Peter the Great. The date of composition is not certain, but we know that it was inspired by the great Catalan victory in Sicily in 1282, and that the author died in 1289. Consequently, it must have been written between 1283 and 1289.

(Munt)   Muntaner, Ramon: *Crònica*   'Chronicle'

Muntaner's Chronicle focuses on the history of the Crown of Aragon, from the conception of King James I (1207) to the coronation of King Alfons the Benign in Zaragoza (1327). It was written from 1325 to 1332. The author Ramon Muntaner served the kings of Aragon as knight and chronicler and intended his work for oral performance.

(Per)   Ramon de Perellós: *Viatge del vescomte Ramon de Perellós i de Roda fet al Purgatori nomenat de Sant Patrici*   'The Journey of Viscount Ramon de Perellós i de Roda to Saint Patrick's Purgatory'

The text, datable to the second half of the 14th and the early 15th centuries, tells the journey of Ramón de Perellós to purgatory, with the aim of visiting King John I.

(Parl)   *Parlaments a les corts catalanes*   'Speeches at the Catalan Courts[6]'

This volume is a compilation of various speeches, delivered at the Catalan Courts between 1355 and 1519.

(EpC)   *Epistolari del Renaixement*   'Collection of Renaissance Letters'

This collection contains letters from between 1478 and 1577. Occurrences of the construction *anar* + Inf appear in one of the merchant Pere Freixe's letters, written to his son in 1506, and in one of his son's letters written to his brother in 1533.

(Ger)   *Cròniques de les germanies: les cròniques valencianes sobre les germanies de Guillem Ramon Català i de Miquel Garcia*   'History of the Germanies:[7] history of the *germanies* of Valencia of Guillem Ramon Català and Miquel Garcia'

The text originates from the first half of the 16th century.

1.3.2.2 Spanish texts in the corpus

The corpus of the study contains the following Spanish texts, which will be referred to by abbreviations assigned to them as follows:

(Cid)   *Cantar de mio Cid*   'The Poem of the Cid', unknown author

This medieval epic poem is the oldest literary work in the Castilian language. It is based on a true story and tells of the brave Castilian hero Rodrigo Díaz de Vivar, known as El Cid. It survived in a manuscript signed by Per Abbad (Abbot Peter). He put the date 1207 after his name, but the existing copy forms part of a codex from the 14th century, kept in the National Library of Spain, in Madrid. Ramón Menéndez Pidal dated a previous text reconstructed by linguistic analysis to circa 1140.

(Mar)   *Vida de Santa María Egipciaca*   'The Life of Saint Mary of Egypt'

This text from the beginning of the 13th century tells of the life of Saint Mary of Egypt, who repented of a life of prostitution and was converted. It was written in Castilian with traits of the Aragonese dialect. It may stem from an Old French text, but the author inserted additional material too.

(Tres)   *Libre dels tres reys d'Orient*   'Book of the Three Kings of the Orient'

The text, written in Castilian dialect, shows features of the Aragonese language. It relates, among other events, the flight into Egypt and the holding of the holy family by the good and the evil thieves. It has been preserved in the same manuscript which also contains 'The Life of Saint Mary of Egypt'.

(GrCo)   *Gran conquista de Ultramar*   'The Great Conquest of Outremer'

This medieval Castilian chronicle was composed probably between 1291 and 1295. It relates the conquest of Jerusalem during the First Crusade, but also includes independent narratives from the French epic. It has been transmitted to us in four manuscripts, the oldest one, kept in the National Library of Madrid, is from 1295. Some critics believe that its author was Alfonso X the Wise, while other sources attribute it to Sancho IV.

(Alf)   *Crónica de Alfonso X*   'Chronicle of Alfonso X'

Alfonso XI commissioned the *Crónica de Alfonso X* in 1344, to memorialize the kingdom of Alfonso X, Sancho IV, and Ferdinand IV. Titles were added later.

(Tris)   *Cuento de Tristán de Leonis*   'The Story of Tristan of Leonis'

This work is one of the medieval Spanish texts which relate the story of Tristan. The late 14th- or early 15th-century manuscript written in Castilian and Aragonese is incomplete, missing the first folios and the end of the story.

(Arn)   *Tratado de amores de Arnalte*   'Treatise Concerning the Loves of Arnalte and
           y Lucenda   Lucenda'

The text is one of Diego de San Piedro's sentimental romances, written in 1491. It contains letters and reflections on love, demonstrating the hopeless frustrations experienced by Arnalte, who repeatedly tries and fails to win the love of Lucenda.

(Pier)   *Pierres y Magalona. La historia*   'Pierres and Magalona: The story of the
           de la linda Magalona y del muy   beautiful Magalona and the very brave
           esforzado caballero Pierres de   knight Pierres of Provence'
           Provenza

The story of Pierres and Magalona is a 15th-century Provençal romance by Bernardo Treviez, which was translated into Spanish in 1519.

(CrPC)   *Crónica popular del Cid*   'Popular Chronicle of El Cid'

This work is one of the chronicles published in Spanish throughout the 16th century, edited in 1512 in Burgos.

1.4 Structure of the book

After the present introductory chapter, the results of the study on the initial stage of the semantic change of the *anar* + Inf construction are discussed in three chapters: the object-scientific part of the research is presented in Chapter 2, while methodological issues are addressed in Chapter 3. Finally, in Chapter 4, I propose a new hypothesis, developed on the basis of the conclusions of this twofold study.

After an introduction to semantic changes of motion verb constructions in the languages of the world, the two first sections of the object-scientific part present the current use of the construction *anar* 'go' + Inf and *ir* 'go' (*a*) + Inf in medieval and current Catalan and Spanish, respectively. Section 2.3 offers a summary of previous literature on the topic, discussing and criticizing the main hypotheses, which raises the necessity of a new approach. After presenting the theoretical background in Section 2.4, Section 2.5 provides a morphological analysis of the occurrences found in the historical corpus. Section 2.6 deals with entrenched local schemas discovered in the use of the motion verb constructions under study and offers a comparison of their use with uses of functionally similar constructions in the same language state. Section 2.7 summarizes the object-scientific results of the research.

Chapter 3, on methodological aspects, opens with an introduction (Section 3.1), then Section 3.2 reviews data sources in pragmatics, and in particular historical pragmatics, while Section 3.3 examines how they are used in research practice. Section 3.4 deals with contextual analysis as a central tool in historical pragmatics methodology, while Section 3.5 offers methods used to guess meanings in historical documents. Section 3.6 focuses on a presentation of the p-model and its notion of *data*, which serves as a basis of the methodological part of the study. Argumentation methods in historical pragmatics and the usefulness of methodological considerations when deciding between rival hypotheses are discussed in this section, too. Relying on the results of the historical study discussed in Chapter 2 and methodological considerations in Chapter 3, I introduce my own position on the semantic change of *anar* + Inf in Chapter 4. I place particular emphasis on the role of pragmatic inferences in the process (4.2) and discuss possible typological parallelisms, as well (4.3).

For the sake of clarity, I present the object-scientific study in a separate chapter, but I refer back continuously to the relevant connected parts in the methodological chapter. Finally, Chapter 5 is a summary which reviews both the object-scientific and methodological results of the study. At the end of this chapter, the new achievements of the study are highlighted in a separate section. Preceding the bibliographic data for the Historical sources and References, there is an Appendix with a list of occurrences of the Catalan constructions *anar* + Inf and *anar a* + Inf found in the texts of my corpus, quoted with a brief context.

CHAPTER 2

Object-scientific part of the study: initial stage of grammaticalization of the Catalan *anar* + Inf construction

2.1 Motivation and research problem

2.1.1 Previous findings on semantic change of motion verb constructions

The motion verb 'go' used in different constructions is the source of several different abstract, grammatical meanings in several languages of the world. In their *World Lexicon of Grammaticalization*, Heine and Kuteva (2002: 155–165) present several grammaticalization pathways which these constructions usually undergo. They offer examples from various languages from a number of different language families where verbs meaning 'go' give rise to andative, change-of-state, consecutive, or continuous (including markers of progressive and durative aspects) markers. Verbs meaning 'go' can be sources of distal demonstratives, habitual markers, hortative suffixes or imperative markers too, and they can develop into markers of allative, future tenses, or purposive meanings in purposive constructions.

Spanish and Catalan, Romance languages close to each other both geographically and linguistically, also use several periphrases based on motion verbs, and similar constructions were already in use in their medieval language stages. For instance, a combination of the motion verb 'go' with the present participle, also known as the gerund, is used to express continuous actions in present-day Spanish and Catalan. The gerund in conjunction with the verb 'go' emphasizes the idea of a continuous, ongoing action, sometimes repetition (see Badia i Margarit 1995: 616–619 for Catalan and Yllera 1999/2000: 3412–3416 for Spanish), as in the following Catalan example:

(1)

L'Antònia em **va**   **anar**   *trucant fins que vaig agafar el telèfon.*
            Aux(go).3Sg.Pres Aux(go).Inf   call.Ger

'Antonia **called me again and again (literally: went calling me)**, until I picked up the phone.'

Combinations of the verb 'go' with the past participle focus on a continuous state of the subject (cf. Yllera 1999/2000: 3424–3428). In this verb phrase the verb 'go' either retains its original meaning or it functions as a periphrasis as in the following Spanish example:

(2)

Los gauchos **iban**   **vestidos**   *de pieles de guanacos.*
            Aux.3Pl.ImpP   dress.Part

'The gauchos **were dressed** in guanaco leather.'

Finally, both Catalan and Spanish use the combination of the verb 'go' and an infinitive, which also existed in medieval Catalan (*anar* + Inf) and in medieval Spanish (*ir (a)* + Inf). The use of the constructions in present-day Catalan and Spanish will be discussed in more detail in Sections 2.1.2.2 and 2.1.2.3, respectively. During the grammaticalization of these periphrases, the verb meaning 'go' lost its full lexical meaning of motion and became an auxiliary. In the Spanish construction, the verb *ir* 'go' grammaticalized into a future marker. In contrast, the Catalan GO-construction evolved into a preterit tense, and it is used to express a perfective past meaning in modern Catalan.

The evolution of all these periphrases presented above, formed with the verb 'go' and a participle in Catalan and Spanish, and also the Spanish purposive construction *ir (a)* + Inf fit cross-linguistically recurrent tendencies of semantic change of motion verb constructions, as documented in Heine and Kuteva (2002). In contrast, the Catalan construction *anar* + Inf does not fit any typical pathway of semantic change. A parallel construction has been documented in Guardiol (a variety of Occitan) with a perfective past value (Jacobs and Kunert 2014), and in other varieties of Occitan, where its use is highly restricted: it is used sporadically as a stylistic variant of the simple past tense (Lafont 1963–1968: 275; Colon 1976/1978; Jacobs and Kunert 2014).

The question arises whether there are similar semantic developments of 'go' beyond the scope of Romance languages. Juge (2008: 29–31, relying on Bybee et al. 1994 and Holmer 1946) lists four possible typological parallels, mentioning four languages in which a verb originally meaning 'go' is used as some type of past marker: Alyawarra, Tukano, Cocama, and Cuna. However, according to the relevant literature on these languages, the use of these GO-constructions is pragmatically marked. Juge (2008) characterizes these possible typological parallels in the following way. In Alywarra, the GO verb does not mark tense in itself and is also apt to express a future meaning. Tukano has a completive GO-construction, which is not restricted to the past tense, since it can also be used in a future context. The discussion of the GO-perfect in Cocama is incomplete. The GO-preterit of Kuna combines a present stem and the suffix *–na*, derived from 'go'. In contrast to the Catalan construction, the GO verb is conjugated in the past, as becomes clear from the brief morphological discussion. In summary, these poorly documented constructions (see also Pérez Saldanya's (1996: 84) references to similar constructions in Cuna, Chadic, Swahili, and Ewe, and Pérez Saldanya and Hualde's (2003: 48) examples from Cuna, Chadic, Swahili, and Tucano) are in stark contrast to various well-documented typological parallelisms of the development of future tenses from purposive constructions (see Heine and Kuteva 2002: 161–163). Thus, the evolution of the Spanish immediate future tense from the GO-construction '*ir* 'go' (*a*) + Inf[1] can be easily explained as a typical representative of a widely attested pathway of semantic change, while the particular evolution of the parallel Catalan construction *anar* 'go' + Inf, untypical for purposive GO-constructions, is more problematic to account for, in the absence of well-documented cross-linguistic analogies.[2]

This volume aims to reveal whether the evolution of the Catalan construction departed from cross-linguistically frequent tendencies right from the beginning and, if so, what the reason is for this. The following sections are a brief digression from our main topic, i.e. the initial stage of grammaticalization of the Catalan *anar* + Inf construction, carried out in order to review the use of 'go' + Inf in current Catalan and Spanish.

2.1.2 *Anar* + Inf in present-day Catalan and *ir a* + Inf in present-day Spanish

2.1.2.1 *The Catalan verb* anar *and the Spanish verb* ir

The Catalan verb *anar* and the Spanish *ir* encode the same concept 'go'. Both the Catalan *anar* and the Spanish *ir* can be used both as main verbs and auxiliaries in the current language state. The morphology of the verb 'go' in these languages is only detailed to the extent necessary to understand the grammaticalization process under study. The formal difference between the infinitives *anar* and *ir* is due to the peculiar history of these Romance verbs. Both are suppletive verbs whose different forms derive from several distinct Latin verbs. In the paradigm of the Catalan *anar* 'go', the stems *va-* (< VADERE) and *an-* (< *ANDARE) can be traced, although in some archaic and dialectal future and conditional forms the stem *ir-* (< IRE) also appears. The paradigm of the Spanish *ir* 'go' contains forms based on three different verbal stems: *i-* (< IRE), *va-* (< VADERE) and *fu-* (< ESSE) (for more details, see Juge 1999 and Moll 1991/2006: 224–225). As a result, the paradigms of the modern Catalan and Spanish verbs meaning 'go' show some formal differences; however, most of the present indicative forms of the full verb and those of the auxiliary are based on the verbal stem *va-* (< VADERE) in both languages, as Table 1 shows.

Table 1. Paradigms of the Catalan *anar* and Spanish *ir* (Juge 2006: 314)

	Catalan auxiliary *anar* and full verb *anar* 'go'		Spanish auxiliary *ir* and full verb *ir* 'go'
	Preterit auxiliary	Full verb 'go'	
1Sg	*vaig (vàreig)*	*vaig*	*voy*
2Sg	*vas (vares)*	*vas*	*vas*
3Sg	*va*	*va*	*va*
1Pl	*vam (vàrem/vem)*	*anem*	*vamos*
2Pl	*vau (vàreu/veu)*	*aneu*	*vais*
3Pl	*van (varen)* [3]	*van*	*van*

In the following, let us consider some present-day Spanish and Catalan data concerning the current use of these GO-constructions in these languages.

2.1.2.2 Anar + *Inf* and anar a + *Inf in present-day Catalan*

The construction *anar* + Inf is used to express a perfective past meaning in modern Catalan:

(3)

Pere II va   *morir l'any 1285.*

     Aux.3Sg.Pres die.Inf

'Pere II **died** in 1285.'

Diachronic research reveals convincingly that the auxiliary of this Catalan periphrastic perfective past has the verb *anar* 'go' as a source. Currently, there are no restrictions on what verbs can appear in the construction.[4] However, in modern Catalan *anar* + Inf the auxiliary *anar* cannot express the original lexical meaning 'go'; the construction is only used as a preterit. In addition, there also exists a simple form used to express preterit meaning. Distribution of the use of synthetic vs. analytic preterit forms varies by areas, dialects, and registers. In general, we can detect a decline in the use of the simple form (Perea 2007; cf. Pérez Saldanya and Hualde 2003: 58), while the analytic version (*anar* + Inf) appears almost exclusively in the spoken language variant of most dialects. In spoken Catalan, the simple preterit is only used in Valencia and some parts of the Balearics (except Menorca), but, except for central regions of Valencia, the periphrastic preterit is also in use in these Catalan-speaking areas. The periphrastic preterit is the preferred form in the rest of the Catalan dialects. In these, the simple past is reserved only for literary purposes, where the alternation of the simple and analytic preterits serves stylistic purposes (Badia i Margarit 1995: 546–548; cf. also Juge 2006). In the Algherese Catalan dialect spoken in the city of Alghero (Sardinia, Italy), the simple past, instead of the periphrastic GO-perfect, has been replaced by the present perfect, possibly due to Italian influence (cf. Scala 2003: 57; Perea 2007).

Although the Catalan verb *anar* evolved into a preterit auxiliary in the construction *anar* + Inf, the verb *anar* 'go' continues to be used as a full lexical verb, as well. Although in modern Catalan the paradigms of the full lexical verb and the auxiliary *anar* almost coincide, they are divergent from each other in some forms, as can be seen in Table 1.[5] Unlike the Spanish *ir a* + Inf construction, the Catalan *anar* cannot express the full lexical meaning 'go' if it is used in the construction *anar* + Inf. In order to indicate a motion with the aim of doing something, the preposition *a* 'to' should be used: *anar a* 'go to' + Inf.[6] In contrast, the verb *anar* 'go' used in the prepositional *anar a* 'go to' + Inf construction can only be interpreted in most cases as expressing the lexical meaning of motion,[7] and not as referring to futurity, past, or any other abstract meanings. In some cases, however, it can be used to express inchoative and conative meanings. Beyond that, in the past tense, where disturbing morphological syncretisms are not present, the construction *anar a* + Inf can be used to express intention, for instance: *Ho **anava a dir**, però vaig callar* ('I wanted to say it, but I remained silent'). It should be noted that the use of *anar a* + Inf in the present tense to convey futurity or intention in modern Catalan is considered a Castellanism and is disallowed in normative language use (Badia i Margarit 1995: 615; cf. Radatz 2003).

However, the construction sometimes expresses futurity or intention also conjugated in the present tense in everyday language use, which may be due to Spanish influence, since the Spanish *ir a* + Inf has similar uses (see 2.1.2.3). Moreover, Radatz (2003) assumes that the use of Catalan *anar a* + Inf with intentional or future meanings is not a Castellanism, but an outgrowth of an indigenous linguistic evolution. He deals with the grammaticalization of the prepositional variant of this Catalan GO-construction and diagnoses that it has reached a stage of semantic change when it is used as a periphrasis meaning 'intention', which was the first step in grammaticalization in the case of the Spanish construction, too. However, the Catalan construction only allows for human subjects. Radatz also reveals indications of the next step, when the intentional meaning evolves into a future tense, and does not interpret them as due to Spanish influence, but rather as a consequence of an internal evolution. Occurrences of *anar a* + Inf expressing futurity can already be detected in Catalan, although alternation of the GO-construction and the simple future is different from their distribution in Spanish; the simple future tense is used in Catalan to express immediate futurity also in contexts where Spanish employs the GO-construction (*ir a* + Inf). Radatz (2003) concludes that the Catalan immediate future tense *anar a* + Inf is just in the process of being born, and although the influence of Spanish favours this process, the main promoting factor is an internal tendency. I think that this problem cannot be clarified unequivocally because of the areal proximity of these languages, and the problem of the 'Catalan immediate future' goes beyond the scope of the present volume.

2.1.2.3 The construction ir a + Inf in present-day Spanish

Throughout this book I will refer to the current Spanish construction *ir a* + Inf as the *immediate future*, following the tradition of Spanish grammars, which mention futurity in the first place among the semantic nuances this periphrasis can convey. However, it can express several other shades of meaning as well, similarly to other morphemes of future tense in the languages of the world. The Spanish immediate future is formed with auxiliary forms that have the verb *ir* 'go' as a source. Currently, there are no restrictions on what infinitives can appear in the construction. In addition to periphrastic use as the immediate future, the verb *ir* retains its full lexical meaning of motion in the construction *ir a* + Inf in some contexts. In these occurrences the verb *ir* should be considered as a main verb, not an auxiliary.

The Spanish *ir a* + Inf can be used to indicate actions that are about to take place shortly, while in other cases the act or event described in the infinitive is conceived as something ongoing, inherent, or as a consequence of a previously made decision (Zamorano Mansilla 2006). The following utterances show the contrast between the simple future (4) and the immediate future (5) in present-day Spanish. While the former makes a general statement, the latter describes the event of dying as inherent and a natural consequence of the situation which exists at the moment of utterance.

(4)

Moriré.

'I will die.' (someday, like everybody)

(5)

Estoy seguro de que **voy**   *a*   *morir.*
                      Aux.1Sg.Pres Prep die.Inf

'I am sure **I am going to die.**' (soon, I feel, for instance because I am ill)

In its grammaticalized uses, beyond conveying futurity, the construction can suggest the emotional involvement of the subject or the speaker in the event, and it is strongly linked to the situation which holds at the moment of utterance (Matte

Bon 1998; Bauhr 1989). Modal meanings can even relegate futurity meaning to the background. Evolution of future markers into modal ones is a common path of semantic change throughout the world. To a limited extent, this periphrasis can also be used as a synonym for the simple future tense. This use is characteristic in spoken language, similarly to the parallel English construction. Zamorano Mansilla (2006) compares the use of *ir a* + Inf with the use of the English *be going to* + Inf construction and finds that they are very similar, the only difference being that the Spanish construction has modal uses too. These modal uses mainly pertain to epistemic modality, but it has some deontic or contrafactual modal uses, too (Bravo Martín 2008; also see Gómez Torrego 1999/2000: 3365).[8] Gómez Torrego (1988: 66–75) attributes the following values to this periphrasis: immediate future, potential modality, inchoative aspect, and intentional modality (see also Gómez Torrego 1999/2000: 3365). Zamorano Mansilla takes into account 500 occurrences of the construction *ir a* + Inf and diagnoses the following distribution of different meanings: 18.4% modal, 13.8% simple future, and 67.8% immediate future meanings. Other grammars of Spanish also mention immediate future as the central use, thus I adopt the traditional term *immediate future* to refer to this construction.

It is difficult to judge whether *ir a* + Inf is really a tense as suggested by the term *immediate future* or simply an aspectual periphrasis. One of the arguments which favours the former opinion is the tendency of *ir a* + Inf to substitute the simple future in spoken language. According to several studies (see Orozco 2007), this analytic form of the future is the preferred way of expressing futurity in different varieties of current Spanish. However, we should note that this phenomenon mainly belongs to spoken language: the contexts where analytic and simple forms of the future are interchangeable without any difference in meaning pertain to the spoken register. Torres Cacoullos (2011: 11) applies the variationist method to discover patterns of usage of *ir a* + Inf. Her study confirms that the overall likelihood that *ir a* + Inf will be chosen has increased dramatically from the 17th to the 20th century in the spoken language. In other words, the periphrastic future has rapidly increased in relative frequency, while the simple future has become the minority variant. Following Sankoff (1988: 154), Torres Cacoullos employs empirical tests to detect differences in meaning, relying on clues in the context of the utterance. These linguistic sub-contexts are represented as factors. Each token is coded for a series of factors for multivariate analysis (e.g. infinitive types entering the periphrasis, adverbs in contexts of use, etc.). Relying on such a multivariate quantitative model, Torres Cacoullos (2011: 11) shows that *ir a* + Inf is favoured: (a) more by dynamic predicates than by stative, perception and psychological verbs, (b) more by the absence of adverbial modification and least by co-occurring indefinite adverbials and, (c) more by interrogatives than by declaratives. Zamorano Mansilla (2006) compares the Spanish construction with the parallel English construction *be going to* + Inf and argues that the Spanish periphrasis expresses future tense and not aspect, since tenses are able to locate a situation in time, relative to the time of utterance, while aspectual periphrases are not. If the construction *ir a* + Inf was an aspectual periphrasis, it would not be able to convey futurity on its own.

Bravo Martín (2008) concludes the opposite: in her opinion *ir a* + Inf is simply an aspectual periphrasis. She maintains that the grammaticalization of the periphrasis from expressing aspect into a future tense has not yet concluded, even though the periphrasis can convey the temporal information of futurity when a temporal adverb carrying this temporal information appears in context. For instance, contexts containing adverbs of posteriority facilitate a temporal interpretation, while other types of contexts favour the older, aspectual interpretation (e.g. subordination introduced by *cada vez que* 'so long as', *siempre que* 'as long as', 'on the supposition that'). Bravo Martín (2008) regards the construction *ir a* + Inf as an aspectual periphrasis, which expresses the prospective aspect. Prospective aspect describes a situation in a chronological relationship with other situations. When using the periphrasis *ir a* + Inf we predicate about an interval of time and indicate the relationship of a given situation with another subsequent situation.

The present volume confines itself to examining only the beginnings of the grammaticalization of the Catalan and Spanish GO-constructions; consequently, it is not relevant here to judge whether *ir a* + Inf expresses prospective aspect or future tense in current Spanish. However, it is worth observing that the phonological reduction in some contexts, namely, the loss of the preposition *a* in the current use of the Spanish construction, shows that the grammaticalization process is still in progress. The result of such a process is a growing formal similarity with the Catalan GO-past. Loss of the preposition *a* is probably possible in Spanish, because the variant without the preposition with a past tense value does not exist in this language, i.e. it does not result in any disturbing syncretism. In the following, let us consider Bravo Martín's (2008) hypothesis on the grammaticalization of the Spanish *ir a* + Inf, which I hope will be useful in the research on the grammaticalization of the Catalan construction.

Bravo Martín (2008: 324–373) characterizes the verb *ir* 'go' as an unaccusative verb with two arguments, which has two different conceptual structures. In the first case, the two arguments of the verb *ir* correspond to the components Object-Theme and Path in the conceptual-semantic structure. This is the case when the Goal of Path is realized as a prepositional phrase as in the following Spanish example (6):

(6)

Pedro va   a   la   playa.
       go.3Sg.Pres Prep Art beach
'Pedro **goes** to the beach.'

However, the verb *ir* 'go' appears in other cases with an infinitive clause introduced by the preposition *a*. Bravo Martín argues that in these cases the infinitive clause does not indicate Place, but Event; in other words, we should not suppose any metaphorical process. Rather, this approach suggests that we should assume another conceptual-semantic structure for the verb *ir*, in which the second argument would be an Event. In contrast to the utterance in (6), which contains a prepositional phrase, in Example (7) there is an infinitive phrase introduced by the preposition *a*:

(7)

Pedro fue   a   visitar   una   exposición.
       go.3Sg.PerfP Prep visit.Inf Art   exhibition
'Pedro **went to visit** an exhibition.'

Some authors do not distinguish between the two cases, but Bravo Martín argues in favour of two different conceptual-semantic structures, and not just one, which then is interpreted metaphorically. In other words, the verb *ir* has two prototypical occurrences in her approach: (a) when the conceptual-semantic argument is a Path, a lexical preposition appears as in (6), and (b) when it is an Event, a functional preposition appears, as in (7). This approach is in accordance with hypotheses of grammaticalization that assume that it is the conceptual-semantic structure that subsists in semantic change in grammaticalization. Bravo Martín opines that the conceptual-semantic structure of the verb *ir* in which the second argument is an Event was the one which subsisted in the evolution of *ir a* + Inf into a future tense. A relation of posteriority between events can be deduced from this structure. This posteriority may also play a role in the grammaticalization of the Catalan *anar* + Inf construction (see Pérez Saldanya's approach in 2.3.2).

It is worth noting, however, that the arguments Path and Event can be realized at the surface structure simultaneously, which suggests that the conceptual-semantic structure of the verb *ir* contains perhaps three arguments, which may, but need not, be lexically realized at the same time:

(8)

$[_{\text{Event}} \text{GO} ([_{\text{Object}}], [_{\text{Path}} ([_{\text{Place}}])] [_{\text{Event}}])]$

According to this analysis, the Event argument remains implicit in example (6)

('Peter goes to the beach (to bathe)'), while in (7) there is an implicit Place argument ('Peter went to visit an exhibition (to the place of the exhibition)'). The realization of these arguments at the surface would usually be redundant, but they can be lexically realized if the intended meaning is not the default interpretation.[9] The supposition of the simultaneous existence of these three arguments in the conceptual-semantic structure of the verb *ir* 'go' allows us to manage all occurrences together. It will be useful when analysing pragmatic inferences in the grammaticalization of the GO-construction.

After considering the current uses of *anar* + Inf and *ir (a)* + Inf, let us now consider how they were used in the Middle Ages.

2.2 The construction 'go' + Inf in the Middle Ages

2.2.1 Introduction

GO-constructions of the type 'go' + Inf were widespread in Romance languages during the Middle Ages. Their use has been documented in the following languages of this period: French (Juge 2008: 48; Colon 1959/1978, 1976/1978; Berchem 1968; Bres and Labeau 2013), Provençal (Occitan) (Juge 2008: 48; Colon 1959/1978, 1976/1978; Berchem 1968; Jacobs and Kunert 2014) and Spanish (Juge 2008: 48). Two varieties of Occitan need a special mention: Gascon (Berchem 1968; Jacobs and Kunert 2014) and Guardiol (dialect spoken in the village of Guardia Piemontese, Calabria, Italy) (Berchem 1968; Jacobs and Kunert 2014). While in the majority of these languages, the use of 'go' + Inf declined during the 15th–16th centuries (cf. Colon 1959/1978, 1976/1978; Berchem 1968: 1163), in Catalan it gives rise to the development of the *go*-past *anar* + Inf. Jacobs and Kunert (2014) document the presence of the *go*-past in two conservative varieties of Occitan as well, namely, Gascon and Guardiol. In Gascon it has a restricted use (still documented in the 1950s), but Guardiol has fully preserved the Occitan *go*-past, which is used in this language variety as the default-preterit for all verbs (Jacobs and Kunert 2014: 190–192).

2.2.2 The construction 'go' + Inf in medieval Romance languages

The Catalan *anar* 'go' + Inf and Spanish *ir* 'go' *(a)* + Inf constructions are already attested in the earliest written documents. The earliest Catalan text in the corpus of the present study is from the 13th century, while the earliest Spanish text is from the 12th–13th centuries, both being the earliest written records in their respective languages. Even a superficial look at occurrences found in our historical corpus suggests that the use of these periphrases shows a great similarity between early Catalan and Spanish. Let us consider the following Catalan (9) and Spanish (10) examples.

(9)

Sobr·assó vengren-li bé·X·cavalers justats e **anaren-lo**   **ferir.**

                                              go.3Pl.PerfP-Clit attack.Inf

'upon this 10 knights on horseback came up to him and **went to attack/attacked** him'

(Desc II 43,28)[10]

(10)

e **fueron-se**   **ferir**   *asi fuerte mente deste primero*

   go.3Pl.PerfP-Refl attack.Inf

encuentro que los cauallos & los caualleros fueron en tierra (a) todos juntos

'at their first encounter, they **went to attack/attacked** each other so ferociously that both the horses and the knights fell to the ground together'

(Tris 20v)

Examples (9) and (10) illustrate typical uses of the medieval Spanish and Catalan GO-constructions and reveal their similarity. Descriptions of battle scenes contain most of the utterances with these GO-periphrases, where they seem to fulfil a function of dynamification, intensifying and drawing the addressees' attention to the event in the story which the speaker considers to be the most important one. *Ferir* 'to attack, to hit, to wound' is the most frequent infinitive in the earliest occurrences of the construction, both in Spanish and Catalan. Moreover, the literature also mentions similar occurrences in Old French and Occitan (cf. Gougenheim 1929/1971; Lafont 1963–1968: 275). Medieval Catalan and Spanish texts contain occurrences of 'go' + Inf with the verb 'go' conjugated in either present or past tense. The similarity of contexts, functions, semantic nuances, and morphological characteristics of the Catalan and the Spanish constructions in medieval texts suggests that they have a common origin.

In the following, let us consider briefly the use of this GO-construction in other medieval Romance languages too, relying on the relevant literature. Among Romance languages it is Occitan in which the situation is most similar to Catalan. The use of the construction in medieval Occitan is discussed in Lafont (1963–1968: 275), in Colon (1976/1978: 146–156), and in Jacobs and Kunert (2014). Lafont (1963–1968: 275) summarizes the history of the Occitan periphrasis in the following way: the earliest attestations with a past tense value date from the late 12th century, much earlier than occurred in Catalan. Its use is so abundant during the 14th and 15th centuries that it competes with the simple past. However, during the 16th and 17th centuries it is attested more rarely, even though it does not disappear completely. Jacobs and Kunert (2014) maintain that the use of the construction has survived in two conservative varieties of Occitan: Gascon and Guardiol. According to the authors, while in Gascon it has a restricted use with only four subtypes (described in Jacobs and Kunert 2014: 183–189 on the basis of Marquèze-Pouey 1955), in the conservative Occitan variety spoken in the village of Guardia Piemontese it functions as the default preterit.

Relying on contemporary grammarians' works, Colon (1976/1978: 147–148) finds that the Old Occitan periphrasis was stigmatized as vulgar, similarly to the parallel Catalan GO-construction. Colon (1976/1978: 146) maintains that the decline in its use was exacerbated by the high prestige of the literary language and the increasing influence of French. Jacobs and Kunert (2014: 182) also propose a causal relationship 'between the weakening of the *go*-past in Occitan and the irresistible advance of the French tongue in southern France, leading to a rapid increase of Occitan:French bilingualism and possible ambiguity between the Occitan *go*-past and the French *go*-future'. The current situation is a consequence of these factors: although sometimes it is replaced by the periphrastic past for stylistic effect, the simple past is predominant in almost all varieties of Occitan, with Guardiol being the only – in the literature often ignored – exception.

Colon (1976/1978) provides some examples of the Old Provençal construction which, in his opinion, convey 'a past tense value', originating from epic and prose. Some texts have Latin parallels, like the homily containing the following occurrence in (11):

(11)

E **va**   *dire*   *Zacharias a l'àngel.*

  go/Aux.3Sg.Pres say.Inf

Et **dixit** *Zacharias ad angelum.*

'And Zacharias **said** (=[11] **?goes to say**) to the angel.'

(Colon 1976/1978: 152)

Colon (1976/1978: 154–156) provides similar examples from Old Gascon too and compares them with fragments from corresponding French and Latin texts. The comparison of texts written in different languages reveals that in places where the Old Gascon uses the GO-periphrasis, the French and Latin texts use the simple past tense (cf. (11)). The term *past tense value* leaves open the question of whether we are dealing with coded meaning or whether the periphrasis expressed some pragmatic or stylistic value in this early language state. Anyway, Jacobs and Kunert (2014: 189) assume that the *go*-past was once a fully productive, grammaticalized preterit in Gascony, which subsisted well beyond the 16th century, when it had disappeared from most other Occitan dialects.

The parallel Old French construction appears in similar contexts and with similar infinitives, as can be observed in the following occurrence from the Song of Roland:

(12)

En mi le camp amdui s'entr'encuntrerent,

Si se   vunt   ferir,   granz colps s'entredunerent

   Refl go.3Pl.Pres attack.Inf

'In the middle of the field they met and they **go to attack/attack/attacked** each other, and they were dealing hard blows to each other.'

(Rol[12] 3567–3568)

Colon (1976/1978) also summarizes literature on the Old French construction and provides examples, as he did in the case of other Romance languages. He emphasizes that there were attestations of the construction in the *chanson de geste* not only in medieval French, but already in Old French, although he notes that these occurrences are not yet instances of a past tense (1976/1978: 144). He regards these uses as an epic tool, which mainly appears in descriptions of battle scenes. Similarly to Catalan and Spanish, the infinitive *ferir* and others conveying analogous meanings are the ones which entered the construction most frequently at that time. Such an occurrence is shown in (13).

(13)

Un coup li va   doner

   blow   go.3Sg.Pres give.Inf

'He **goes to give/gives/gave** him a blow.'

(Boeve 631, Colon 1976/1978: 144)

According to Colon (1976/1978: 141), the phrase *va dire* 'goes to say/(she/he) says/said' was used in Old French to introduce a direct quotation, too. This observation should be compared to our observations about the Catalan and Spanish constructions given in sections 2.6.3.2 (subsection h) and 2.6.4.3.

The use of the French GO-construction declined at the beginning of the 17th century. Colon (1976/1978: 144) does not indicate the exact meaning of the Old French construction, only mentioning that it is yet not a tense, merely an epic tool. It is strange that although the Old French and Provençal constructions seem to be very similar, Colon attributes a different meaning to them. In any case, occurrences cited in Colon (1976/1978) seem to be parallel in all the Romance languages he treats: Catalan, Spanish, Provençal, and French. Given that the stylistic replacement of the simple past with the periphrasis in Occitan was usual even in Colon's time, perhaps he projected this later use to early Provençal occurrences.

Finally, it is interesting to note that contemporary French has a similar usage of 'go' + Inf with a narrative interpretation, which emerged around the mid-20th century, as described in Bres and Labeau (2013). However, the authors emphasize that while the old construction is combined with past tenses, the new one is used in present and future contexts. They argue that the contemporary construction does not derive from the old one, but from the future reading of the *go*-periphrasis, used as a metanarrative construction.

2.2.3 Shared grammaticalization?

The comparison presented in the previous section suggests that although early occurrences of the GO-construction in various Romance languages cannot be unequivocally considered examples of a past tense, they are still indications that these constructions started to follow a path of grammaticalization towards evolving into a past tense, even if this process did not reach its end. Moreover, these constructions were in use over centuries, in French until the 17th century, in Occitan from the end of 12th to the 16th–17th centuries, and sporadic occurrences appear even later. The evolution of these constructions may exemplify the path of grammaticalization followed by the Catalan construction, which raises the possibility of *shared grammaticalization*.[13] The question is whether these periphrases have common origins or developed independently from each other in these neighbouring languages. It is strange that this obvious question rarely appears in the literature on the topic (but see Jacobs and Kunert 2014).

According to the definition provided by Robbeets and Cuyckens (2013b: 1), shared grammaticalization is 'a state whereby two or more languages have the source and the target of a grammaticalization process in common'. The authors note that the targets may not have reached the same degree of grammaticalization. As for the grammaticalization process presented in this book, shared grammaticalization is possible only as regards the medieval language states of Western Romance languages, when Catalan had this construction in common with other languages. As for the modern language state, Catalan shares this feature only with Guardiol.

Shared grammaticalization can be attributed to different reasons: universal principles of language change, language contact, formal coincidence with contact, or common ancestorship. The main concern of the studies compiled in the book on shared grammaticalization edited by Robbeets and Cuyckens (2013a) is how we can identify which one of these reasons played a role in actual cases of grammaticalization. Shared grammaticalization is also possible in cases in which there is no geographical or historical connection between two or more languages, since these processes are driven by universal principles of grammatical change. Thus, some studies in the volume deal with criteria allowing us to exclude universal principles as a motivation for shared grammaticalization. The chapter by Heine and Nomachi (Robbeets and Cuyckens 2013a) concentrates on the phenomenon of 'rare grammaticalization', while Robbeets's contribution examines 'grammaticalization clusters'. Grammaticalization processes which are not widespread cross-linguistically, but which are rather shared by a group of languages in a certain geographical area, should not be accounted for by universal principles of grammatical change. Similarly, we should exclude universal motivation when a cluster of shared developments appears in a particular area, while they usually do not correlate cross-linguistically. In these cases, the motivation may be an areal or genealogical-historical connection between the languages concerned.

As for the grammaticalization process discussed in the present volume, at first sight it looks like an instance of rare grammaticalization, thus its parallel presence in neighbouring languages could be accounted for by language contact, and not by universal principles of linguistic change. However, I have to note that although the development of the Catalan construction as a whole is due to the particularities of the history and linguistic structure of Catalan (e.g. the decline of the simple past tense) – in other words, the overall process of grammaticalization instantiated by the Catalan *anar* + Inf is not widespread cross-linguistically – some of its stages or sub-processes can be linked to cross-linguistically attested changes.

Returning to the comparison of the uses of different Romance periphrases, our conclusions may be the following: the analysis of surviving historical texts suggests that the earliest attestations of the GO-construction under analysis appeared in Occitan, and neighbouring languages, i.e. French, Catalan, and later Spanish, might have taken it from that language. Consequently, this grammaticalization process may instantiate shared grammaticalization by language contact. This exactly is the hypothesis that is defended by Jacobs and Kunert (2014), who argue that the *go*-preterit developed first in Occitan and from there diffused into Catalan-speaking areas. They (Jacobs and Kunert 2014: 194) lay out the hypothesis with the following words: '[W]e do not share the assumption (often implicit in the literature on the *go*-past) that the construction developed *within* Catalan. Rather … we think that the *go*-past developed first in Occitan and only subsequently diffused into Catalan'. While I can accept the main idea and their argumentation, I think we should not interpret this statement literally. Historical data show that Catalan did not adopt a fully developed *go*-past. We could rather say that Catalan adopted an early usage of the construction, perhaps through the adoption of some frequently used phrases, and then, the past meaning could develop in a parallel way in these two neighbouring languages, due to the continuous and vital language contact.[14] The early contexts of use in Catalan do not show a fully developed *go*-past, and the gradual semantic change can be revealed in Catalan data. The most important idea in this respect could be what Jacobs and Kunert emphasize in the following lines:

> [I]t is doubtful that Old Occitan and Old Catalan represented two distinct languages in the modern sense of the word. In line with e.g. Radatz (2011),

we are more inclined to assume that in the late Middle Ages they formed a fluid Catalo-Occitan dialect continuum … Consequently, we can hypothesize that the *go*-past simply diffused (or 'travelled') from the Occitan to the Catalan end of the dialect continuum, as a type of 'travelling isogloss'.

(Jacobs and Kunert 2014: 197–198)

However, it is worth considering another possibility: whether grammaticalization has occurred independently in each of the languages mentioned above, and that the observed shared grammaticalization may therefore be attributed to universal principles of language change. I think that such an account would be difficult to prove; due to the geographical connectedness of the languages concerned (Catalan, Spanish, Occitan, and French), it would be complicated, if not impossible, to exclude the possibility of common influence. Universal motivation can also be excluded, on the basis of criteria provided by Heine and Nomachi, and Robbeets (see relevant chapters of Robbeets and Cuyckens 2013a). The only reason why I consider this possibility here is that Colon (1976/1978), interestingly, accepts it (cf. the section about polygenesis, Colon 1976/1978: 170–171). However, his argumentation is not entirely convincing. Colon, after examining the use of the GO-construction in medieval Catalan, Provençal, and French, asks the question: are we dealing with polygenesis or monogenesis? He highlights the fact that although the necessary conditions were given in every Western Romance language, this tendency to language change can be attested only in Gallo-Romance languages. According to Colon, historical occurrences suggest that the construction developed in an epic context, but they do not make it possible to focus on one given language, from which it might have spread to others. In other words, monogenesis can be excluded. Citing Henrichsen, he also notes that the construction appeared in Old Provençal and in Old Catalan at the same time. Although Colon indicates an earlier date than Henrichsen, he acknowledges that we do not have Catalan narrative texts concerning the beginnings of any such change and also that the Provençal examples are sporadic and their dating is uncertain. He does not consider French in his argumentation. In summary, Colon excludes the hypothesis of monogenesis on the basis of a lack of historical sources and evidence. This argumentation seems to be surprising, since in the absence of sufficient data we cannot argue in favour of a hypothesis, nor against it.

It should be added that Colon (1976/1978: 140) mentions scholars who treat the use of the construction in French as a Gasconism,[15] that is, they treat the phenomenon as an example of shared grammaticalization by language contact. In a footnote he considers this a mistaken supposition, but does not explain the reason there, only discussing polygenesis later on (Colon 1976/1978: 170–171). In sum, I do not find Colon's argumentation against monogenesis convincing.

The dialectal distribution of the use of the periphrasis versus the simple past can be taken into account as evidence when considering the possibility of shared grammaticalization. The simple past is currently used in the spoken language of the central and southern regions of Valencia, and some regions of the Balearics (except Menorca), but in every case in alternation with the periphrastic form. In the variety of Catalan spoken in the city of Alghero (Sardinia, Italy), the simple past has been replaced by forms of the present perfect and not the construction *anar* + Inf, perhaps under the influence of Italian (cf. Scala 2003: 57; Perea 2007). All these are Catalan dialects geographically distant from Occitan, and this distribution favours the hypothesis according to which the construction formed in the Occitan language and its use later spread to Catalan.[16] Jacobs and Kunert (2014: 197) also argue in favour of this hypothesis. They draw attention to the fact that the lack of the *go*-past in Alghero 'contrasts crucially with the data from Guardia Piemontese, the village that was settled in roughly the same period as Alghero, but whose speech is characterized by a fully productive *go*-past'. Unfortunately, we do not have written sources for Guardiol until the late 19th century (Jacobs and Kunert 2014: 191), Catalan remaining the only language that provides data for historical research on the beginnings of the development of this *go*-past.

In sum, it seems to be likely that we are looking at a case of shared grammaticalization by language contact. However, this must be stated with some reservations, because we do not have enough evidence, given that historical documents from the early stage of this evolution are missing. Evidence which favours this hypothesis comes from the geographical distribution of the use of the periphrasis under study and from the observation that the phenomenon concerned has not been detected in other languages of the world (except for some partially similar but incompletely documented sporadic cases reported in Section 2.1.1). Nevertheless, this latter reason argues against a shared grammaticalization attributed to universal principles at the same time. As mentioned above, relying on data based on written texts we can also hypothesize that the GO-construction first appeared in Occitan, and the neighbouring languages (French, Catalan, and later Spanish) acquired it later under the influence of that language. In any case, the use of the construction declined in most of the Romance languages during the 16th–17th centuries, Catalan being the only language which currently uses it as a full past tense.[17]

2.3 Findings in previous research on the formation of the Catalan GO-preterit

2.3.1 Hypotheses on the formation of the Catalan preterit *anar* + Inf

The main strands in the literature concerning semantic change in the grammaticalization of the Catalan GO-preterit (*anar* + Inf) are the following. Some scholars base their analyses primarily on medieval present tense occurrences of *anar* + Inf and derive the current Catalan *anar* + Inf construction from a historical present usage (Colon 1959/1978, 1976/1978; Badia i Margarit 1981; Pérez Saldanya 1996; Pérez Saldanya and Hualde 2003), while others argue that present tense occurrences do not show characteristics of the historical present, but are rather already grammaticalized forms (Bruguera 1981; Juge 2002, 2006, 2008). Another hypothesis relates the origins of the Catalan GO-preterit to the inchoative aspect (cf. Detges 2004; for a criticism of the inchoative hypothesis, see Colon 1959/1978; Nagy C. 2008a, 2010b), although this account is consistent with the historical present hypothesis. Finally, some authors examine the semantic change of *anar* + Inf from a cognitive point of view and start from the deictic nature of the verb *anar* 'go' (Calvo Pérez 1995; Martos 1996). Let us discuss these hypotheses in more detail.

2.3.2 The historical present hypothesis

Since the auxiliary forms of the perfective past *anar* + Inf in modern Catalan formally seem to be in the present tense, it is not surprising that some authors base their analyses on present tense occurrences of *anar* + Inf and derive the current Catalan *anar* + Inf construction from them. Such an analysis seems to be obvious at first sight. Because these authors regard present tense occurrences of the verb *anar* as instances of a historical present usage, I will refer to their approach as the *historical present hypothesis*.

The historical present hypothesis is present, implicitly or explicitly, in several works on the Catalan GO-preterit. Lafont (1963–1968: 276) already made a link between the historical present and the formation of the Catalan GO-preterit. However, the more detailed version of this account is provided in Colon (1959/1978) and (1976/1978). Colon's analysis is one of the most thorough studies on this past tense, based on a wide range of data, not only from Old Catalan but also from other Romance languages. I will give a summary of the development of *anar* + Inf step by step, as provided in Colon (1959/1978) and (1976/1978). In the first stage, *anar* + Inf is a periphrasis with a motion verb conjugated as the context requires, used in order to animate the narration. The periphrasis with a verb conjugated in the present tense does not only occur in a preterit context, but also in general descriptions in the present as a stylistic tool (see Colon's example 1959/1978: 127–128). That is why Mendeloff (1968) assumes the existence of two different *anar* + Inf constructions: one of them a perfective form and the other an inchoative periphrastic present. Berchem (1968) also supposes that the construction primarily served to express the present tense, and the past tense meaning appeared as a secondary function. Berchem derives the current use of the construction from an inchoative present, but he does not offer any details of the semantic change process. In contrast, Colon focused not on the inchoative present occurrences, but on present tense occurrences which appear in a past context of narrative passages. He interprets such occurrences as examples of the historical present. He opines that the preterit forms of the construction disappear, relegated to the background by the narrative present usage. The GO-periphrasis with an auxiliary in the historical present later acquires a past tense value. According to Colon (1959/1978; 1976/1978: 156), we cannot examine the beginnings of the history of the Catalan *anar* + Inf, because it already appears with a past tense value in the earliest texts. Colon (1959/1978, 1976/1978) observes that early occurrences of the periphrasis highlight the most crucial event in a narration, which usually has serious consequences.[18] Colon attributes a 'past tense value' to such examples. Similarly, Soldevila (1963–1968) states that occurrences of the periphrasis in Muntaner's Chronicle are representatives of a periphrastic past tense, and that Muntaner uses the periphrastic perfective past for a stylistic reason, in order to break the monotony and lend variety to his style. He observes that in Muntaner's Chronicle, more and more occurrences of the construction *anar* + Inf appear towards the end of the text, and he explains this – constructing a parallel with his own language use – by claiming that the author became enthusiastic about the use of the construction; in other words, he contributes accumulated occurrences in the same context to Muntaner's individual style. However, he simply quotes the occurrences and does not argue in favour of a past tense meaning for them.

Colon's analysis is based on a wide range of data not only from Old Catalan, but also from other Romance languages. However, his account also contains an inconsistency between two of his statements, namely, (a) that the periphrasis *anar* + Inf already has a past tense value in early texts (Colon 1959/1978: 120, 122), and (b) that we are dealing with a historical present use (Colon 1959/1978: 128–129). This inconsistency results from the circumstance that both claims are based on the same data sources, that is, the same historical texts representing the same language state. It is obvious that the two statements, inasmuch as they concern the same language state, cannot be true simultaneously. If the first statement can be maintained, the GO-verb should already function as a grammatical marker, or at least be grammaticalized to some degree. However, this would contradict the statement that we are dealing with a historical present usage, because in this case the GO-verb should refer to real motion.

We can solve this inconsistency in two ways. (a) We can suppose that Colon intends the two statements to refer to two different language states. In this sense, the second one concerns an earlier period of the history of the Catalan, from which we do not have documents, as Colon also points out. Given the lack of direct sources from this language period, it is not possible to support this claim properly. (b) We can consider the term 'past tense value' as referring to an implicit meaning, and not to a coded meaning. Colon does not indicate which one of these two possibilities he accepts.

Colon's statement, according to which occurrences of *anar* + Inf would already have a past tense value in early texts, is debatable if we interpret 'past tense value' as a coded meaning and not as a general value of the historical present. Analysis of historical data suggests that we are not dealing with fully grammaticalized forms in these early cases. In contrast with Colon's claim, on the basis of corpus data it can be claimed that the distribution of the periphrasis is highly constrained in the earliest texts: the range of contexts in which the periphrasis can appear is very limited, only certain infinitive types can enter the construction, and its frequency is very low in comparison with its current use. In summary, it is hard to understand what Colon means by past tense value, because he does not yet use the terminology of (historical) pragmatics. If we do not interpret his term as referring to a coded meaning, his opinion can be maintained, since a more thorough analysis of historical occurrences in their contexts suggests that a past tense reference is really already present in the early occurrences as implicit meaning. If we interpret the term *past tense value* in such a way, statements (a) and (b) turn out to be simultaneously maintainable.

Segura-Llopes (2012) agrees with Colon's suppositions, and he adds further evidence in their favour. He studies the evolution of the Catalan GO-past while participating in the ongoing project of the development of the Electronic Corpus for the Old Catalan Grammar (Corpus Informatitzat de la Gramàtica del Català Antic [CIGCA]), and thus he is working with a large corpus. Occurrences of *anar* + Inf with the verb *anar* conjugated in the present tense are analysed as historical present use in his approach, as well. However, he associates the development of the meaning of perfectivity with occurrences containing the verb *anar* in the preterit.

The historical present hypothesis is accepted also by Pérez Saldanya (1996),[19] who analyses the occurrences of *anar* in the present tense as instances of a historical present use. His work is important, because he offers a detailed analysis of the actual steps involved in the semantic change in the grammaticalization of *anar* 'go' + Inf, based on data from Bernat Desclot's Chronicle. He describes semantic change in a cognitive grammatical framework as a metaphorical transfer of some semantic aspects belonging to the cognitive domain of SPACE into the more abstract domain of TIME. Between the two he supposes a narrative-aspectual domain; in other words, he believes that at a certain stage of its evolution, the verb *anar* was an 'aspectual auxiliary/semi-auxiliary'. He notes, as Colon (1959/1978: 120) had already observed, that the Catalan periphrasis under consideration shows a preference for narrative genres and is absent in others, such as theology, philosophy, or didactic works. Narrative texts are characterized by the use of verbs in the third person and conjugated in the past, and the description of event sequences usually reflects their chronological order. Pérez Saldanya claims that the verb *anar* occurs in these texts conjugated in typical narrative tenses, namely, in the perfective past and in the historical present. He assumes that the properties of these tenses played a crucial role in the grammaticalization of *anar* + Inf. He hypothesizes that, in an intermediate stage of this process, the construction became a narrative-aspectual periphrasis, in which the verb *anar* was used as a marker of discourse-coherence, sequentiality, and the significance of the event described in the infinitive. Pérez Saldanya (1996: 86–89) highlights the pivotal role of the following factors in the grammaticalization of the construction: (a) the goal of the movement is not indicated by means of a prepositional phrase, and (b) the narrative context allows for an inference that the action expressed by the infinitive was accomplished.

Pérez Saldanya (1996: 88–91) explains the development of the construction in a perfective past tense relying on the following factors: (a) temporal and aspectual characteristics of tenses (perfective past, historical present) in which the verb *anar* appears; (b) implications and inferences allowed by the narrative context; (c) lexical aspect of the infinitives entering the construction (punctual events). He thinks that the reason for this evolution may be found in various formal and structural problems that the simple past presented in medieval Catalan. A detailed criticism of this approach is provided in Juge (2006: 327–333).

In contrast to the approach adopted by Colon, who associates the semantic change in the construction with historical present occurrences, Pérez Saldanya, when revealing the semantic change in the construction, takes into account both variants: occurrences with the verb *anar* in the perfective past, and occurrences in the historical present (at least he believes this to be the case). He claims that at an intermediate stage, *anar* + Inf was a narrative-aspectual construction, and at that stage both present and past auxiliaries were allowed. Occurrences with the auxiliary in the past tense disappeared at a later stage, when the verb *anar* was used as a past tense auxiliary, conjugated exclusively in the present. This hypothesis constitutes a transition from Colon's (1959/1978) to Juge's (2006) approach. This latter rejects the supposition according to which there would be a historical present usage in these places and explains the formally present tense occurrences in a different way (see 2.3.6). Pérez Saldanya's hypothesis can be reconciled with Juge's (2006) account, if we emphasize that, in the description of the semantic change, they both start from occurrences with the verb *anar* conjugated in the perfective past.

The above presentation of Pérez Saldanya's approach already indicates that the grammaticalization of the Catalan *anar* 'go' + Inf is not only a question of tense, but also of aspect, in two senses. On the one hand, the periphrasis evolved into a perfective past in Catalan, unlike in other Romance languages which have a similar periphrasis with a future meaning. Badia i Margarit (1995: 646–647) characterizes the current Catalan *anar* + Inf as a tense which indicates perfective aspect and describes punctual events.[20] Where can this perfective meaning come from? Describing the way future markers evolve in the languages of the world, Bybee et al. (1994: 268) point out that, in cases when a new future marker develops, 'the overt or inherent aspect of the construction is progressive, present, or imperfective'. The earliest medieval Catalan *anar* 'go' + Inf occurrences, in which the GO-verb is in a perfective past, do not fit this schema. This fact may explain why this construction did not evolve into a future tense. The aspect of infinitives entering the periphrasis may also have had a role in the semantic development of the whole construction (for more details, see Colon 1959/1978; Pérez Saldanya 1996; Juge 2008; Nagy C. 2008a). On the other hand, we need to discuss the problem of aspect because many authors agree that, at an intermediate stage in its grammaticalization into a perfective past tense, the construction was used as an aspectual periphrasis (cf. Pérez Saldanya 1996; Detges 2004). We have already seen Pérez Saldanya's claim that, at a certain stage of its semantic change, the *anar* + Inf construction was a narrative-aspectual periphrasis. Pérez Saldanya (1996: 84) produces as evidence the observation that both hypotactic and paratactic structures appear in the same function in similar contexts, as can be observed in the following texts:

(14a)

*E quant foren tots entrats, ajustaren-se en .I. loch, e puys **anaren** a avant e **feriren** en la gran pressa dels sarraÿns.*

'when they all had entered, they gathered together in one place, then they **went** forward **and attacked** in a great multitude of Saracens'

(Desc II 135)

(14b)

*encontraren-se ab la host dels sarraÿns e **anaren ferir** en éls*

'they met the troops of the Saracens and **went to attack/attacked** them

(Desc II 103,15)

In the literature about the Catalan GO-construction, we can find another hypothesis concerning its aspectual value, which I call the *inchoative hypothesis* and discuss in the following section.

2.3.3 The inchoative hypothesis

The inchoative hypothesis consists in the supposition that early occurrences of the Catalan *anar* + Inf represent the inchoative aspect. Motion verb constructions are appropriate to express the inchoative aspect because the beginning of the act that is considered to be the aim of the motion and the end of the motion overlap in the real world. Several instances of the development of inchoative aspect markers from verbs denoting some kind of movement have been documented in the world's languages (see Detges's 2004: 214 examples in n. 22), thus it is not surprising that it also appears in the literature on the grammaticalization of the Catalan *anar* + Inf construction. The inchoative hypothesis has already been discussed in Colon (1976/1978: 156–163), who presents this view as a widely accepted one and cites Meyer-Lübke (1890–1902: 357) as its source. However, Colon does not accept it. His central counter-argument is that there are several instances (not only in Catalan, but in French too), which combine the GO-periphrasis with the inchoative verb 'begin' (Catalan *va començar* + Inf 'began' + Inf/?'begin to begin' + Inf). This goes against the hypothesis that *anar* 'go' + Inf in itself expresses beginning. In no way is it my intention to defend the inchoative hypothesis; however, I note that, although this criticism seems to be convincing at first sight, it can easily be rejected. We know that grammaticalization is a process which unfolds gradually, during which forms belonging to different stages of grammaticalization can coexist. It can occur that the construction under study expresses aspectual (in this case inchoative) meaning in some contexts of a certain language state, while in others the verb *anar* can be conceived of as a preterit tense auxiliary. In this sense the occurrence *va començar* '(she/he) began' can be analysed as an example of a fully grammaticalized analytic past tense.

Colon (1976/1978: 159) assigns to *anar* 'go' + Inf the function of animating the storytelling, and not that of referring to the beginning of events: 'After its concrete semantic content of a verb of motion was bleached, the verb *anar*, due to its dynamic character, only has the function of animating the narration and not that of indicating the beginning or the end of an action chronologically' (my translation).[21] Pérez Saldanya and Hualde (2003: 47), however, do not accept this, arguing that it is difficult to conceptualize how a perfective meaning can be derived from an inchoative meaning.

The inchoative hypothesis reappears in Detges (2004: 213–215), where it is claimed that the *anar* 'go' + Inf in Old Catalan was an aspectual periphrasis meaning 'abrupt start'. Detges (2004) supposes that GO-constructions in various medieval Romance languages are representatives of the inchoative aspect. He offers a very detailed analysis of the semantic development of the Catalan GO-construction, analysing the process in several steps which he considers to be independent from each other, and emphasizing the role of discursive factors in the process. In his scenario, the inchoative aspect is the first stage in the development of the Catalan *anar* 'go' + Inf. It expresses the beginning of an action or process, thus it can be paraphrased as 'begin to' + Inf. Inchoative constructions are compatible with durative verbs, which denote actions or processes that can be characterized by a time interval. The inchoative aspect signals the left starting point of this interval (Kiefer 2006: 169–170). As Detges (2004) describes it, the inchoative aspect suggests cognitive relevance, and its use with such an aim is a very common strategy in various languages around the world. Detges (2004: 214) mentions several instances of the development of inchoative aspect markers from verbs denoting some kind of movement.[22] He claims that the *anar* 'go' + Inf construction is appropriate to express the inchoative aspect because the end of the motion and the beginning of the act considered to be the aim of the motion overlap in the real world, and its use instantiates the conceptual setting from MOVEMENT to BEGINNING (Detges 2004: 213). Due to this overlap, the motion directs the attention to the proposed action itself. As described in Detges (2004), the inchoative aspect is a considerable discourse-structuring device in the world's languages. The discourse-structuring nature of the Catalan GO-construction, conceived of as inchoative by Detges, is the basis of the following step in its development, inasmuch as it serves as a rhetorical tool for emphasizing dramatic moments of a story (cf. also Pérez Saldanya 1996: 76). However, the first step of the analysis is the most important one, because if this is not correct, the analysis should be modified.

Detges's hypothesis is connected with the historical present hypothesis in the sense that he claims that the present auxiliary variants of *anar* + Inf are representatives of a historical present usage. In Detges's scenario, the next step in the semantic change of the Catalan construction was that the present tense variant conventionalized as a foreground-marker, as a consequence of its high frequency (Detges 2004: 218): it highlighted the event conceived as the most important one in a narrative segment. Detges (2004: 218) maintains that 'the present-tense variant was used more frequently at all times', but he does not support the high frequency of the present auxiliary version with (quantitative) data. Moreover, statistical data suggest the opposite hypothesis, according to which the preterit auxiliary variant of *anar* + Inf was more frequent at the beginning and was displaced by the present auxiliary forms only later (see Section 2.5.3; Bruguera 1981; Juge 2008: 38). The high frequency of the present tense variant seems to be rather a consequence of semantic change than a motive for it.

Finally, Harris (1982) also accepts the historical present and inchoative hypotheses, although he deals with the history of the Catalan *anar* 'go' + Inf construction only briefly. In the following let us consider a supposition which can be linked to the historical present hypothesis and appears as a subsidiary hypothesis in several authors.

2.3.4 The 'historical present as stronger version' hypothesis

As mentioned above, the use of the construction with the verb *anar* conjugated in the present tense is also detected in a preterit context in medieval Catalan documents. As we have seen in Section 2.3.2, such examples are conceived of as testimonies of a historical present usage in the opinion of some scholars. Whatever the case may be, this is the variant which extends its use and displaces the preterit forms, and this needs an explanation. An interesting hypothesis has been proposed to explain this replacement. According to Colon (1959/1978: 129), it is perfectly understandable that the auxiliary was conjugated in the historical present instead of the preterit, because their function is the same and both 'unite their virtues'. Colon does not explain why they should be used together if they have the same function, and why it is not enough to use historical present forms or the periphrasis independently in order to gain the effect of dynamism. Segura-Llopes (2012: 142–143) also assumes that the historical present and the periphrasis (analysed by him as a demarcative construction) have similar roles and they intensify each other. The construction formed in this way is more expressive, in his opinion. Similarly to Colon, Segura-Llopes (2012: 142) also considers that this higher expressivity was the reason why the use of the present tense variant was later extended.

Detges (2004: 218) also maintains that the version which has the auxiliary in the present is the 'stronger alternative', which combines two effects, namely, a 'dynamification effect' and, due to the historical present, a 'presentification'. At the same time, he notes that the historical present is not indispensable in order to achieve the 'effect' of the periphrasis; the tense of the auxiliary is optional. Detges does not explain why the historical present would appear at all, and why only in the case of this periphrasis, but not in other verbal forms.

A similar supposition can be revealed in Pérez Saldanya and Hualde (2003: 55), when they explain that the use of the present in a past context emphasizes the dynamism implied by the use of the periphrasis. In those cases, they assume a 'stronger grammaticalization'; it would, however, be problematic to explain what is meant by this term.

It would be hard to judge what effect different versions of the periphrasis had on contemporary addressees. I argue that if a stronger and a weaker version of the periphrasis actually existed in the medieval language, their difference should be reflected in the documented uses. However, contexts in historical documents do not reveal a 'weaker version vs. stronger version' opposition: the present and the past tense versions appear in early texts in the same types of contexts and with the same infinitives.

2.3.5 The deixis hypothesis

Another account of the grammaticalization of the Catalan *anar* + Inf construction takes as a starting point the notion of deixis. Deictic reference occurs whenever we integrate in the interpretation pieces of contextual information concerning spatial, temporal, or interpersonal aspects of the speech event (see Tátrai 2010). Using the verb 'go' is a case of spatial deixis, and as Tátrai (2010: 225) notes, some deictic expressions that represent spatial relations can function metaphorically as temporal deixis. Calvo Pérez (1995: 270–273) analyses the history of the Catalan *anar* + Inf construction on the basis of the deictic nature of the verb 'go' and emphasizes that it can be used not only for spatial, but also for temporal deixis. The verb 'go' in Catalan indicates the subject's movement in the space towards a certain place (except for the place where the addressee is) or the motion of some other entity away from the speaker. Accordingly, the periphrasis formed with the verb 'go' can express the speaker's or some entity's movement towards the future, but also their movement away from the speaker, towards the past. Calvo high-lights that the deixis is based on the perspective 'here and now' of the speaker (for egocentric deixis, see Tátrai 2010: 226–227), and he argues that this fact explains why the conjugated verb of the construction is in the present and why the past forms disappeared. If this is correct, it is hard to understand why they existed at all, and this account leaves many historical occurrences without any explanation. In the historical texts, there are many occurrences with the verb 'go' conjugated in the past, and they seem to be the earlier variants, that is, their primordial role is unquestionable. Pérez Saldanya (1996: 78) objects to this approach in that it assumes an intermediate stage (present perfect or immediate past tense) which is not documented in the historical texts of either of the languages which used the periphrasis. Most of the occurrences I have analysed originated from medi-eval chronicles, and this narrative genre is characterized by use of the preterit; the present perfect is not a characteristic tense in this genre. However, in the *Epistolari del Renaixement* there is a very interesting occurrence of *anar* + Inf (15), which seems to reflect this intermediate stage. This occurrence is surprising because this text postdates the chronicles:

(15)

per què us sia avís que digau a vostra sogra que, si ella sabia l'enuig que madò Riquera n'hauria si sabia lo que en Vallcanera diu, jo crec ella ne pendria la mort o una malaltia on perdria lo seny. Per tal, vostra mare vos prega digau a vostra sogra que li faça una lletra e que no li'n diga cosa niguna … a bé que nosaltres la

vam   ***pregar*** *que no li'n diga cosa ninguna, e així ho <u>ha promés</u>, no sé qué*
go.1Pl.Pres   ask.Inf

es farà. Vós ja la coneixeu e la que us feia quan hic éreu; e de ço us prega vostra mare que ho digau a vostra sogra.

'and so I warn you, tell your mother-in-law how resentful Mrs Riquera would be if she knew what Vallcanera says about that, I think she would die of it or it would make her sick and drive her crazy. That is why your mother asks you, tell your mother-in-law to write to her, but not to say anything about that … although we **have asked** her <u>not to say</u> anything to her about that, and she <u>has promised,</u> I don't know what will happen. You know her and what she did with you when you were there; and that is what your mother asks you to tell your mother-in-law.' (the merchant Pere Freixe to his son, Bartomeu, 4 June 1506.)

(Ep2 145)

In this fragment from a letter, the form *diga* is conjugated in the present subjunctive in the subordinated clause following the clause containing the periphrasis (*que no li'n diga cosa ninguna* '<u>not to say</u> anything to her about that'), which indicates that the form *vam pregar* cannot be a preterit verb. If it was, then in accordance with the agreement between the tenses of verbs in related clauses, the verb 'say' should be conjugated in the past subjunctive. The fragment *la vam pregar que no li'n diga cosa ninguna* in itself could be interpreted as a plan in the future ('we are going to ask her not to say anything to her about that'), but this interpretation is excluded by the subsequent part of the letter. The following verbal form (*ho ha promés* 'she has promised') is conjugated in the present perfect; in consequence, the act of asking was performed, because the person who was asked has promised to do the requested thing. On the basis of all of these pieces of information, it can be concluded that the verbal form *vam pregar* is used to convey past tense reference, which cannot be a perfective past, only an immediate past. The frequency of this usage and its development cannot be reconstructed on the basis of this single occurrence.

Martos (1996) presents an account which can also be connected to the deixis hypothesis. He also starts from the conceptual structure of the verb 'go'. He explains that with Catalan's peculiar conception of time the construction 'go' + Inf evolved into a past tense in Catalan, but into a future in other Romance languages. Relying on Lakoff and Johnson (1986), Martos differentiates between two ways of conceptualizing time, both starting from space: (a) time is a moving object which moves towards us or (b) time is a fixed location with respect to which we are moving (towards the future). He points out that Lakoff and Johnson erroneously generalize the direction of the motion of time as a motion towards the future, relying on their own perception of the world. He characterizes Catalan as a regressive language (in opposition to progressive ones), which is looking from the present to events already realized (Martos 1996: 117). Martos confirms his hypothesis with other linguistic data which also reflect the regressivity of the Catalan understanding of time and concludes that the inchoative meaning is also directed towards the past in this language. Martos's hypothesis is very interesting; however, if it were correct, it would be hard to explain the ongoing development of an immediate future in Catalan, which some authors suppose (see Radatz 2003). Moreover, while there is no doubt that the evolution of the Catalan GO-past can be explained in part by the peculiarities of the Catalan language, a hypothesis which relies mainly on processes also detected in other languages of the world would be much more persuasive.

Such an account is presented in Bourdin (2008), who treats together the grammaticalization process in Catalan and some cases of semantic change in certain varieties of spoken Arabic, examining the grammaticalization of 'come' and 'go' into markers of textual connectivity.[23] He notes that the Catalan GO-construction followed the same trajectory, what is more, it arrived to the end of this grammaticalization cline, since the verb 'go' developed into a past tense auxiliary in it. He accepts the account presented by Detges (2004). In some languages the verb 'go' is employed to encode suddenness or counter-consecutiveness (Bourdin 2008: 45), which Bourdin relates to the deicticity of this verb, namely, that it refers to a motion towards a point other than the deictic centre. Bourdin (2008) represents the semantics of 'go' and 'come' by the following schema: in the case of 'go', the element 'directed motion' is combined with 'otherness', while in the case of 'come', it is combined with 'identity'. The verb 'go' is appropriate to evolve into a marker of textual connectivity which encodes counter-consecution due to the presence of the component of 'otherness' in its semantics: it expresses that the narrative takes a turn other than expected. The typical grammaticalization cline is instantiated by 'come' and 'go' in Kera, where these verbs have developed into markers of consecution and counter-consecution, respectively (see examples provided in Bourdin 2008: 38). Bourdin (2008: 51) explains these grammaticalization processes as instances of metaphorical mapping from space to time. In his opinion, a narrative text constitutes a structured space, where the storyline moves along a main path, made up of successive segments of the story.

There are two paradoxes related to the deicticity of 'come' and 'go'. First, it seems that there are no grammaticalization clines travelled exclusively by 'come' or exclusively by 'go'. Second, 'come' and 'go' may be grammaticalized into future or past tense markers, expressing sequentiality in an intermediary stage of the change (Bourdin 2008: 51–52). It may be possible to solve these paradoxes indicated by Bourdin by a deep morphological and contextual analysis of the actual historical developments, as we can see in this book concerning the second paradox: 'go' provided a marker of the past tense in Catalan, while it developed into a marker of the future tense in Spanish; however, these changes took place in different morphological environments.

Although they provide some interesting suggestions, approaches concentrating on deixis lack contextual analysis due to their theoretical background. Semantic change in grammaticalization can always be linked to special contexts, and their description is very important, in order to understand how the process works. An approach which relies strongly on a detailed analysis of contexts can be complemented with a cognitive approach. The relation between metaphorical and metonymic processes in semantic change in grammaticalization is discussed in Hopper and Traugott (1993: 77–87), in Kearns (2002), and in Dér (2008: 24–26), among others. By metonymic processes what is generally meant is the dynamism of pragmatic inferencing in context. References to these two kinds of processes complement each other in the description of meaning change in grammaticalization, depending on our research perspective (see Hopper and Traugott 1993: 86–87; Dér 2008: 25). Pragmatic inferencing in the actual contexts of language use (which motivates semantic change according to some approaches) may lead to changes also characterizable as metaphorical mappings, operating on a larger time scale; in the case of the verb 'go' it is a metaphorical transfer from SPACE to TIME.

2.3.6 The preterit hypothesis

Another group of authors rejects the supposition that Catalan narrative texts containing early occurrences of the periphrasis are characterized by the use of the historical present. The most prominent representative of this approach is Juge (2006), who claims that forms in the current paradigm of the Catalan perfective past *anar* + Inf apparently in the present tense resulted from analogy, and this process only took place after the past meaning of the whole construction had already consolidated. Consequently, the morphologically present forms are not examples of the historical present, but fully grammaticalized forms. Juge's (2002, 2006, 2008) works on this topic are significant, because he examines the morphological aspect of the grammaticalization of *anar* + Inf more thoroughly than previous accounts. The author bases his hypothesis on the observation that semantic change has already occurred in some examples with the verb *anar* in the preterit. I will therefore use the term *preterit hypothesis* to refer to this account, opposing it to the historical present hypothesis. Juge (2006: 319) presents corpus data in support of his claim: in his corpus 32 of the 157 (20.4%) instances of *anà* 'he went' 'are plausibly cases of the periphrastic preterit' and, similarly, there are 164 instances of *anaren* 'they went' in all, with 40 (24.4%) of these 'being good candidates for the periphrasis'. In summary, he claims that in his corpus every fourth or fifth instance of the GO-construction can be analysed as a grammaticalized form.

As mentioned above, Juge bases his claims on an exhaustive morphological analysis of data originating from a historical corpus. The present–preterit syncretism of 1Pl and 2Pl forms of *anar* 'go' in Old Catalan plays a crucial role in his approach. In other words, present and preterit are formally identical in the first and second plural indicative (*anam* 'we go/we are going/we went', *anats* 'you go/you are going/you went') in this language state, as is shown in Table 2. Juge (2006: 320) claims that 'such forms, especially the 1Pl, are abnormally frequent because they are first person narratives'. The reinterpretation of *anar* + Inf as a periphrasis with a present tense auxiliary was possible due to the high frequency of the 1Pl form *anam*. As a consequence, the use of the present extended to the rest of the paradigm of the auxiliary. Later, new forms were created from the root *va-* by analogy with the regular preterit forms (cf. the paradigms in Table 1).

Juge claims that it is the frequency of 1Pl forms which triggers this process. In his corpus, 209 of 346 occurrences represent the 1Pl form *anam* 'we go/we are

Table 2. Paradigm of the verb *anar* in Old Catalan (Juge 2006: 320)

	Present and preterit indicative of *anar* 'go' in Old Catalan	
	Present	**Preterit**
1Sg	*vaig*	*ané*
2Sg	*vas*	*anast*
3Sg	*va*	*anà*
1Pl	*anam*	*anam*
2Pl	*anats*	*anats*
3Pl	*van*	*anaren*

going/we went', which is a relatively high frequency. If I take into account only occurrences of the periphrasis in the indicative, in my corpus I have found a relatively high frequency of the morphologically ambiguous 1Pl form *anam* only in the Chronicle of James I (while the form *anats* was completely missing). This text contains 24 syncretic forms, in contrast to 19 preterit and only 2 present forms, i.e. over half of the occurrences are ambiguous between past and present tenses. However, 1Pl was more sporadic in other texts, if it was used at all. Among the 235 occurrences found in Muntaner's Chronicle there are only 6 ambiguous *anam* forms, while in the other texts there are none. Pérez Saldanya and Hualde's (2003: 55–56) objection to Juge's account is that most of the grammaticalized occurrences are in the third person singular, and, furthermore, the verb 'go' also appears conjugated in the present tense in other medieval Romance languages where the syncretism observed in Old Catalan is not present. As to the first criticism, I think it is relevant to note that frequency data given at this point of Juge's argumentation concern all occurrences of the verb *anar*, not only those in the periphrasis *anar* + Inf. This is reasonable, because forms of *anar* occurring in periphrasis as well as those which occur independently can influence what paradigm is evoked for language users during linguistic interaction. Whatever the case may be, it would be worth testing whether frequency data will remain the same when investigating this phenomenon in a larger corpus. Pérez Saldanya and Hualde's second criticism also requires careful consideration. As for occurrences in other languages, an examination should be made of the quantity and text type in which they are present. Beyond Catalan, I have taken into account only occurrences from medieval Spanish texts, but even the comparison of data from these two languages has revealed several differences. In the Spanish texts of my corpus I have detected a lower proportion of occurrences conjugated in the present tense, and their distribution was different from what it was in Catalan (see 2.5.3). Present tense forms of *anar* mainly have occurred in the *Cantar de mio Cid* ('Song of my Cid'), a representative of the medieval Spanish epic style, where the verb 'go' appears in the periphrasis conjugated in different tenses, as is usual in this genre (cf. Szertics 1981). In other words, we are dealing with a different genre from that of the Catalan chronicles which are most frequently examined in the literature, and which represent the narrative genre. As for Spanish narrative texts, the use of the preterit dominated, or the use of verbal tenses was different from what it was in Catalan narrative texts (for more details see Section 2.5.3).

Before Juge, Bruguera (1981) had already argued against the historical present hypothesis, and it was he who drew attention to the ambiguity concerning verbal tense in the case of 1Pl and 2Pl forms. He also raised the question of whether these verbal forms were in the historical present. Other difficulties originate from the fact that some historical documents do not use punctuation. For instance, in medieval Catalan manuscripts which do not mark word stress, there is a formal coincidence between the present and preterit tense of the 3Sg forms of first conjugation verbs. For instance, both the forms *parla* '(he) speaks/is speaking' and the end-stressed *parlà* '(he) spoke' can appear in medieval texts written as *parla* (see Bruguera 1981: 31). The present and preterit tense of the 3Pl forms also coincide formally in every conjugation. The context of the occurrence can sometimes be of help: in a description using the present tense, where other unequivocally present verbal forms also appear nearby, we tend to interpret the form *parla* as the present tense. In contrast, in the case of a narrative text that relates past events, the decision is more complicated, since the narrative present and the preterit are both typical in this type of text. In order to decide how to interpret the form *parla* in such a text, we have to take into account our knowledge about the use of the narrative present, and also the place of the occurrence of *parla* in the text: in other words, the decision depends on whether we expect a historical present form in the place in question or not. Bruguera examines data from the Chronicle of James I and he rejects the claim according to which it would be a historical present use in these places in the text. Instances of the form in question from the Chronicle of James I are interpreted as perfective past verbs in Bruguera (1981). Bruguera argues that beyond these morphologically ambiguous forms there are no other verbs conjugated in the present in the relevant contexts of the text. There are only two instances of *anar* + Inf in this chronicle which contain the verb *anar* conjugated unequivocally in the present (this quantity contrasts with 48 past occurrences); however, they can also be interpreted as instances of the analytic GO-preterit. Bruguera essentially accepts Colon's account as presented above, but he misses the analysis of occurrences from the Chronicle of James I. He notes that although the three Catalan chronicles were composed in almost the same period, that of James I contains only two forms conjugated in present tense, while the others feature many more. He thinks this text has an archaic character, also reflected in other linguistic features. This fact is relevant here, because this text contains the occurrences missed by Colon, which can be used to reveal the beginnings of the grammaticalization of the Catalan GO-past. Among others, this is the reason why I have included this text in my corpus.

Juge (2006: 319) provides various pieces of evidence to show why instances of *anar* conjugated in the present tense in this periphrasis cannot be interpreted as examples of the historical present. In a later work Juge (2008: 33–37) offers a thorough analysis of the historical present as well, and concludes that the prototypical characteristics are the following: present tense verbs are used with past temporal reference to express foregrounded information; present progressive or past verbs express background information; and there are extended sequences of present tense verbs. The use of *anar* + Inf in Old Catalan does not show these characteristics. The texts examined by Juge (*Llibre dels fets* and Desclot's *Crònica*) do not contain any occurrences of *anar* conjugated in the present tense and used as a main verb to mark the past. Another piece of evidence against the historical present usage in these chronicles comes from the observation that they show almost no use of present tense verbs with past temporal reference. It would be strange to suppose a historical present usage which concerns only occurrences of a single construction, namely, *anar* + Inf.

2.3.7 Summary: hypotheses in the previous literature

Let me now summarize the main hypotheses on the grammaticalization of the Catalan *anar* + Inf. It is clear that the most important differences consist in (a) which one of the two morphological variants (namely, the past or present tense variant) is hypothesized to be affected by the semantic change, and (b) what meaning is attributed to the construction at an intermediate state of the semantic change:

A. The historical present hypothesis

Colon (1959/1978) and (1976/1978):

anar (historical present/preterit) + Inf 'motion with the aim of doing something' → *anar* (preterit) + Inf relegated to the background → *anar* (historical present) + Inf → <u>semantic change</u> → *anar* (auxiliary formally in present tense) + Inf 'preterit tense'

Pérez Saldanya (1996) and Pérez Saldanya and Hualde (2003):

anar (historical present/preterit) + Inf 'motion with the aim of doing something' → <u>semantic change</u> → *anar* (present/preterit) + Inf 'narrative-aspectual periphrasis' → *anar* (preterit) + Inf relegated to the background → <u>semantic change</u> → *anar* (auxiliary formally in present tense) + Inf 'preterit tense'

B. The inchoative hypothesis

Detges (2004):

anar (historical present/preterit) + Inf 'motion with the aim of doing something' →
<u>semantic change</u> → *anar* (present/preterit) + Inf 'inchoative periphrasis' → <u>seman-
tic change</u> → *anar* (preterit) + Inf relegated to the background → *anar* (present) + Inf
'foreground-marker' → <u>semantic change</u> *anar* (auxiliary formally in present tense) + Inf
'preterit tense'

C. The deixis hypothesis

Calvo Pérez (1995) and Martos (1996):

anar (historical present) + Inf 'motion with the aim of doing something' → <u>semantic
change</u>, metaphorical mapping from SPACE to TIME → *anar* (auxiliary formally in
present tense) + Inf 'preterit tense'

D. The preterit hypothesis

Juge (2002, 2006, 2008):

anar (preterit) + Inf 'motion with the aim of doing something' → <u>semantic change</u> →
anar (auxiliary formally in preterit tense) + Inf 'preterit tense' → morphological changes
in the auxiliary by analogy → *anar* (auxiliary formally in present tense) + Inf 'preterit
tense'

2.3.8 A brief syntactic overview: reanalysis[24]

Although discussion of syntactic aspects of the grammaticalization of *anar* + Inf
lies outside the subject of the present volume, I consider it relevant to offer a
brief overview of syntactic analyses in accordance with the hypotheses presented
above. I summarize different possibilities of the reanalysis of *anar* + Inf, starting
with the account proposed by Pérez Saldanya (1996: 98–100), who deals with the
syntactic aspects of the grammaticalization of the periphrasis, too. He differentiates between three phases of reanalysis ((i) *anar* as a motion verb, (ii) *anar* as an
aspectual auxiliary, and (iii) *anar* as a temporal auxiliary):

(i)

$[_S$ Lo comte$_i$ $[_{VP}$ anà $[_S$ PRO$_i$ ferir los enemics]]] 'The count went to attack the enemies.'

$[_S$ Lo comte$_i$ $[_{VP}$ va $[_S$ PRO$_i$ ferir los enemics]]] 'The count goes to attack the enemies.'

(ii)

$[_S$ Lo comte$_i$ $[_{VP\,Asp}$ anà $[_{VP}$ ferir los enemics]]] 'The count went to attack the enemies.'

$[_S$ Lo comte$_i$ $[_{VP\,Asp}$ va $[_{VP}$ ferir los enemics]]] 'The count goes to attack the enemies.'

(iii)

$[_S$ Lo comte$_i$ $[_{VP\,Temp}$ va $[_{VP}$ ferir los enemics]]] 'The count attacked the enemies.'

In the first phase the verb *anar* functions as a main verb. In the second phase the construction is morphosyntactically more constrained: it can be used only in a narrative context with the verb *anar* conjugated either in the present or the preterit tense. After reanalysis has taken place, the construction turns out to be morphosyntactically even more restricted. In the third phase it can be conjugated only in the present tense, although the selection restrictions of the verb *anar* disappear, which leads to a higher frequency (from about 1350). Detges's analysis would be similar from a syntactic point of view, the only difference being that he would attribute a different aspectual meaning to the periphrasis in the second phase.

The syntactic analysis consistent with Colon's (1959/1978) approach would be different:

(i)

$[_S$ Lo comte$_i$ $[_{VP}$ anà $[_S$ PRO$_i$ ferir los enemics]]] 'The count went to attack the enemies.'

$[_S$ Lo comte$_i$ $[_{VP}$ va $[_S$ PRO$_i$ ferir los enemics]]] 'The count goes to attack the enemies.'

(ii)

$[_S$ Lo comte$_i$ $[_{VP\,Temp}$ va $[_{VP}$ ferir los enemics]]] 'The count attacked the enemies.'

$[_S$ Lo comte$_i$ $[_{VP\,Temp}$ anà $[_{VP}$ ferir los enemics]]] 'The count attacked the enemies.'

(iii)

$[_S$ Lo comte$_i$ $[_{VP\,Temp}$ va $[_{VP}$ ferir los enemics]]] 'The count attacked the enemies.'

Colon's account is different in the respect that both present and past auxiliaries can be used in the phase in which the periphrasis already expresses a temporal meaning. Present auxiliaries later disappear from use.

In contrast, only preterit auxiliaries would be possible in the second phase, according to Juge's (2006) analysis. We should also consider only the preterit forms in the first phase, since in his approach present tense variants should be interpreted as grammaticalized forms and correspond to the third phase:

(i)

$[_S$ Lo comte$_i$ $[_{VP}$ anà $[_S$ PRO$_i$ ferir los enemics]]] 'The count went to attack the enemies.'

(ii)

$[_S$ Lo comte$_i$ $[_{VP\,Temp}$ anà $[_{VP}$ ferir los enemics]]] 'The count attacked the enemies.'

(iii)

$[_S$ Lo comte$_i$ $[_{VP\,Temp}$ va $[_{VP}$ ferir los enemics]]] 'The count attacked the enemies.'

In summary, we can see that different accounts of the semantic change in the grammaticalization of *anar* + Inf provided in the relevant literature would entail different syntactic analyses.

2.4 Theoretical background and research questions

2.4.1 The historical pragmatic approach to grammaticalization

After discussing the relevant literature, in the present section I will present the research I have carried out in the framework of *grammaticalization theory*.[25] Hopper and Traugott's 'cline of grammaticality' reflects a widely accepted definition of grammaticalization: a lexical or content item advances from a lexical status to a grammatical one or from a less grammatical status to a more grammatical one, as represented in (16).

(16)
> *content item > grammatical word > clitic > inflexional affix*

> (Hopper and Traugott 1993: 7; cf. also Meillet 1912/1921, cited in Alonso 1982: 194; Pons
> Bordería and Ruiz Gurillo 2001: 317)

This scale reflects the loss of a lexical meaning and the gaining of a more abstract one but does not tell us where the latter comes from. In current approaches to grammaticalization, in addition to formal considerations, increasing attention has been paid to the pragmatic environment in which this process occurs. Since grammaticalization processes take place in specific contexts, they can be described in pragmatic terms. Traugott's (2003: 645) new definition of grammaticalization can serve as a starting point for pragmatic studies on the topic. She defines it as 'the process whereby lexical material in highly constrained pragmatic and morphosyntactic contexts is assigned grammatical function, and once grammatical, is assigned increasingly grammatical, operator-like function'.

My research aims to provide a comprehensive analysis of the initial stage of grammaticalization of the Catalan *anar* 'go' + Inf construction, investigating it from a historical pragmatic perspective. From a historical pragmatic point of view, it is essential to examine mechanisms of everyday language use, since interaction between language users, the speakers' communicative strategies, and the hearers' pragmatic inferencing in particular contexts play a crucial role in language change. Since semantic change in grammaticalization requires specific contexts to take place, a pragmatic model of semantic change should put a special emphasis on context and take into account the whole utterance, not only isolated lexical units. It should consider the relation between coded meaning and context, and the interaction between conceptual structure and context in semantic change. Historical pragmaticians obtain most pieces of information about contexts in which semantic change emerges by relying on historical documents. Research practice suggests that the process of grammaticalization can be reconstructed relying on written sources, through the examination of historical utterances in contexts and the differentiation between types of contexts which are associable with successive stages in grammaticalization. Contexts where semantic change occurs are usually characterized by semantic ambiguity, or they trigger inference mechanisms. In the following section I discuss the role pragmatic inferencing plays in semantic change in grammaticalization.

2.4.2 Pragmatic inferences in semantic change in grammaticalization

A thorough investigation of the semantic aspects of grammaticalization has led to a subtler notion of grammaticalization which recognizes that it is not only about a loss of meaning, but about enrichment of meaning as well (cf. Sweetser 1988, cited in Heine et al. 1991: 110; Dér 2008: 19–30). Traugott (2003: 632) notes that, even if there is a loss of concrete lexical meaning, there is no loss of semantic complexity. She points out that, in addition to semantic losses, there are also gains (cf. Sweetser's 1988 'loss-and-gain model', which claims that some of the original, often relatively concrete semantic components of a lexeme may be generalized or even lost, but more abstract ones may be gained, as well as new pragmatic meanings). Pragmatic inferences emerging during language use can also be sources of this kind of semantic gain. Within the field of historical pragmatics, the role of pragmatic factors in grammaticalization has recently been the focus of considerable attention and the subject of several studies. These studies emphasize that the regularity of semantic change cannot be accounted for without paying attention to the implicatures and inferences that arise in language use and later attach to a certain construction by a process of conventionalization in the course of language change (Hopper and Traugott 1993; Traugott 1999; Levinson 2000; Kearns 2002; Traugott and Dasher 2002/2004).

The coded meaning of lexical items and constructions can be changed in the course of time by the influence of the context of use, and grammaticalization involves a process of conventionalization of particularized conversational implicatures (Hopper and Traugott 1993; Traugott 1999; Levinson 2000; Kearns 2002; Traugott and Dasher 2002/2004). At an early stage of this process, the grammaticalizing construction may induce associations and inferences which later attach to it in a context in which it is frequently used (Hopper and Traugott 1993: 81–84; Traugott 2003: 634–635). However, the exact steps in the process have not been adequately studied and are not yet well understood. First Grice (1975/1989) and later Levinson (2000) alluded to the possibility of the generalization and subsequent conventionalization of particular conversational implicatures, but they did not describe the process in detail. Traugott (1999) adds that conventional implicatures can become semanticized (cf. Traugott and Dasher 2002/2004: 35). The scale in (17) summarizes how pragmatic meaning from context becomes semantic:

(17)

> *particularized conversational implicature → generalized conversational implicature → conventionalized implicature → semantic/coded meaning*
>
> (Grice 1975/1989; Levinson 2000; Traugott 1999; Traugott and Dasher 2002/2004)

In the original Gricean definition, particularized conversational implicature is carried 'on a particular occasion in virtue of special features of the context', while generalized conversational implicature is carried 'normally, in the absence of special circumstances' (Grice 1975/1989: 37). From the point of view of meaning change, generalized conversational implicature is the most relevant stage because it indicates the first step in the meaning change process, when the content of the implicature is beginning to be context-independent. Generalized conversational implicatures are still cancellable. However, identifying this stage is far from obvious, partly due to the fact that distinguishing between different kinds of implicatures is also problematic. 'Cancelability', a particularity of generalized conversational implicatures, follows from their above-mentioned non-ad hoc nature: they are default inferences, but they can be cancelled in a particular case in a context where the implicature disappears as a result of adding subsequent utterances. Grice describes two ways of cancelling:

> It may be explicitly canceled, by the addition of a clause that states or implies that the speaker has opted out, or it may be contextually canceled, if the form of utterance that usually carries it is used in a context that makes it clear that the speaker is opting out.
>
> (Grice 1975/1989: 39)

At the same time, Grice points out that cancelability is a necessary but not sufficient property of conversational implicatures, i.e. also of generalized conversational implicatures:

> Now I think that all conversational implicatures are cancelable, but unfortunately one cannot regard the fulfillment of a cancelability test as decisively establishing the presence of a conversational implicature.
>
> (Grice 1978/1989: 44)

Grice (1975/1989: 37) already notices that it is easy to treat a generalized conversational implicature as if it were a conventional implicature and he also offers some criteria for differentiating between them. Although he emphasizes that cancelability does not prove the presence of a generalized conversational implicature, this criterion remains useful in obtaining a negative result: if it is not fulfilled it means that we do not have a generalized conversational implicature in the particular context.

The possibility of cancelling implicatures later has been called into question by some authors (see Tsohatzidis 1994 and Nemesi 2006: 35–36). The failure of cancelling raises the problem that the content under consideration should be considered a conventional implicature, because this is the type of implicature which cannot be cancelled. However, the property of cancelability as provided in Grice can also be interpreted in the following way: there must be a context where cancellation is possible, but the content does not have to be cancellable in every context.[26] In other words, if we can find a context where the content considered to be a conversational implicature turns out to be uncancellable, this does not mean automatically that it is not a conversational implicature; it only shows that the distinction between conversational and conventional implicatures raises several difficulties. Whatever the case may be, cancelability shows whether a meaning content is, or is not yet, context-dependent. In historical research it is even more problematic to determine whether a given content is cancellable or not: it is necessary to consider several contexts belonging to the same language state (cf. Diewald 2002). In order to diagnose whether a meaning is context-dependent or not, it should be examined whether there is any context in the given language state where the content does not appear. If there is, the content can be taken as cancellable. The criterion of cancelability is understood in the present research in such a way.

In a previous paper (Nagy C. 2010a) I examined the role implicatures play in semantic change in grammaticalization through a historical pragmatics study. The main theoretical finding of this study was that implicatures play a double role in semantic change in grammaticalization. On the one hand, inferential contents that first appear as particularized conversational implicatures can become part of the semantic meaning of a certain form, attached to a certain construction as a new semantic component. On the other hand, implicatures can also promote the salience of a certain meaning component within the meaning of a lexeme and influence its conceptual reorganization.

As is clear from the above considerations, when dealing with the role of pragmatic inferencing in language change, the relevant literature primarily attempts to reveal the role conversational implicatures play in the process, relying on the Gricean theory. However, the terms *conversational implicature* and *pragmatic inference* are often used as synonyms. Distinguishing between the notions of *implicature* and *pragmatic inference* is too complicated to be exhaustively treated here, so I discuss only some aspects of this problem. It is sufficient to quote the widely known book on grammaticalization by Hopper and Traugott (1993) as an example of the conflated use of the two notions. Hopper and Traugott's conflated use of the terms *pragmatic inference* and *implicature* is a result of the fact that they describe the role pragmatic inferencing has in grammaticalization processes relying on the Gricean model,[27] and at the same time they try to bring together approaches focusing on either the speaker or the hearer (Hopper and Traugott 1993: 64–67). Among motivations for change they discuss communicative strategies, based on general cognitive processes, employed by speakers and hearers in their interactions, but emphasizing that the change is not goal directed. They mention the possibility that the hearer processes the speaker's communicative act in ways that do not match the speaker's intentions (Hopper and Traugott 1993: 64), but they do not distinguish between pragmatic inference and implicature. The Gricean model, however, defines implicature as dependent on the speaker's intention. Since the authors apparently use the terms *implicature* and *pragmatic inference* as synonyms, or at least they do not say explicitly what difference they see between the two, we should suppose that they use the term *implicature* in a broader sense than the Gricean approach. Hopper and Traugott describe the mechanism of implicature in semantic change as a metonymic process, and they interpret it in contrast to metaphor. A discussion of their relationship and a possible distinction between implicature, metonymic, and metaphorical processes is provided in Kearns (2002), who summarizes several aspects of this problem concerning semantic change discussed in the literature.

In order to avoid misunderstanding, throughout this book I will not use the terms *implicature* and *pragmatic inference* as synonyms. Regarding the reasons for this use, Kearns (2002: 2) quotes Levinson (1983: 127), who explains that, according to Grice's approach, *implicature* is a general term which is applied in opposition to *what is said*, and as such it can be understood as a term for any kind of non-truth-conditional inference. Given that the Gricean model is a well-known approach, I prefer to take from it the original notion of conversational implicature, as dependent on the speaker's intention. In the original Gricean framework, implicature, beyond the fact that it is opposed to *what is said*, is dependent on the speaker's intention. When processing utterances, the hearer's inferencing does not necessarily match the speaker's intentions: the hearer may reach also interpretations not intended by the speaker. Consequently, the notion of conversational implicature is not a suitable one to fully describe the hearer's inferences. Neither is it absolutely assured that the addressee understands properly all contents intended by the speaker. Thus it can be supposed that contents inferred but not intended by the speaker, which consequently cannot be regarded as implicatures, may play a role in semantic change.

Further theoretical considerations also confirm the need to distinguish the notions of implicature and pragmatic inference. A discussion of intentionality and speakers' and hearers' perspectives in various forms of language use provided in Németh T. (2015) reveals the possibility of differentiating between particularized conversational implicatures and other inferences drawn by the hearer. The author emphasizes that perspective and intentionality are closely related notions, since speakers can have intentions only within a particular perspective, and in order to infer someone's intention we have to take her/his perspective. According to Németh T. (2015: 69–70) the relationship between perspective and intentions can be used to differentiate between particularized conversational implicatures and other types of inferences. As mentioned above, the notion of implicature is dependent on the speaker's intentions. The speaker's intention to imply a certain piece of information for the hearer can be understood as an integrated part of her/his perspective. The addressee can infer the speaker's intention to imply something, and the implicated piece of information itself, only by adopting her/his perspective, otherwise s/he cannot achieve this. Németh T. mentions two ways of drawing other types of inferences: the hearer may infer pieces of information not intended by the speaker, either from her/his egocentric perspective, or from a perspective that s/he assumes wrongly to be that of the speaker. These three kinds of informational content, their relationship, and the role they play in semantic change need further investigation.

In accordance with the above considerations, what I mean by implicature is an inference dependent on the speaker's intentions, while the term pragmatic inference covers all kinds of not-truth-conditional inferences in which pragmatic factors play a role, independently from whether the speaker intended it to convey them or not. In this sense, pragmatic inference is a broader term than implicature; conversational implicature is conceived of as a special kind of pragmatic inference. The exclusive use of the term *pragmatic inference* would also lead to misunderstandings, since in the discussion of the relevant literature we should not simply exchange the term *implicature* for *pragmatic inference*. Throughout this book I will therefore use both *implicature* and *pragmatic inference*, in the senses explicated above.

2.4.3 Research questions

After presenting the theoretical background, in the following section I discuss the research on the grammaticalization of the Catalan *anar* + Inf construction. My aim has been to answer the following research questions when analysing occurrences found in the historical corpus.[28]

   i.   What is the distribution of the occurrences of 'go' + Inf with and without the preposition *a*, and how does it change over time?

   ii.   What is the distribution of occurrences with the verb 'go' conjugated in the present and the past tenses and how does it change over time?

   iii.   Is there any difference in meaning between the morphological variants?

   iv.   What meaning did the Catalan *anar* + Inf construction express at intermediate stages of its grammaticalization?

> v.   What similarities and differences can be revealed between uses of the Catalan and Spanish GO-constructions and other periphrases used in similar contexts in the Middle Ages? What conclusions can be drawn from these similarities and differences?
>
> vi.   Are the present tense variants of 'go' + Inf instances of use of the historical present, and if not, what are they?
>
> vii.   Should the present or the past tense variant be our starting point when describing the semantic change in the construction?
>
> viii. How and in what contexts may semantic change have been initiated?
>
> ix.   Why did the Catalan construction follow a different evolutionary path from that followed by similar motion verb constructions in other languages around the world?

The analysis of the historical corpus raises the problem that we do not have direct access to native language competence concerning earlier language stages, and researchers have only substitute competence of the language state they are studying (see Forgács 1993–1994). How can our research be objective in such circumstances? Here and now, in the object-scientific sections, I follow the traditional methods of presenting the results of my research, without disturbing the reader with the methodological difficulties they imply. When it was possible, I took into consideration formal clues in the interpretation of the historical occurrences that will be discussed in Section 3.5. Section 3.6.2.2 provides a discussion of methodological difficulties in historical research and of the reliability of data originating from historical corpora which are analysed by relying on substitute competence.

2.5 The construction 'go' + Inf in the historical corpus: morphological questions

2.5.1 Introduction

In this section I discuss the first three research questions, which concern the morphology of the Catalan *anar* + Inf and Spanish *ir (a)* + Inf constructions. I repeat them here for convenience:

> i.   What is the distribution of the occurrences of 'go' + Inf with and without the preposition *a*, and how does it change over time?
>
> ii.   What is the distribution of occurrences with the verb 'go' conjugated in the present and the past tenses and how does it change over time?
>
> iii.   Is there any difference in meaning between the morphological variants?

2.5.2 Occurrences with preposition *a*

The following tables show the distribution of occurrences of the Catalan *anar* + Inf (Tables 3a and 3b) and the Spanish *ir (a)* + Inf (Tables 4a and 4b) in the texts of the corpus, according to the tense of the verb 'go'. In the interest of clarity, texts are presented in chronological order and the distributions of variants with or without the preposition *a* are shown in separate tables.

Table 3a. Distribution of occurrences of the Catalan construction *anar* + Inf in the corpus

	Jau	Desc	Munt	Per	Parl	EpC	Ger	
Preterit	19	69	71	1	1	-	-	161
Present/preterit syncretism	24	-	6	-	-	-	-	30
Present	2	27	157	13	-	1	-	200
Total	45	96	234	14	1	1	-	391

Table 3b. Distribution of occurrences of the Catalan construction *anar a* + Inf in the corpus

	Jau	Desc	Munt	Per	Parl	EpC	Ger	
Preterit	-	2	6	-	-	1	27	36
Present/preterit syncretism	-	-	-	-	-	-	-	-
Present	-	-	-	-	-	-	-	-
Total	-	2	6	-	-	1	27	36

Table 4a. Distribution of occurrences of the Spanish construction *ir* + Inf in the corpus

	Cid	Mar	GrCo	Alf	Tris	Arn	Pier	CrPC	
Preterit	4	-	143	21	41	-	-	51	260
Present	18	5	-	1	11	-	-	2	37
Total	22	5	143	22	52	-	-	53	297

Table 4b. Distribution of occurrences of the Spanish construction *ir a* + Inf in the corpus

	Cid	Mar	GrCo	Alf	Tris	Arn	Pier	CrPC	
Preterit	-	-	16	14	11	-	16	27	84
Present	-	3	-	1	1	1	-	-	6
Total	-	3	16	15	12	1	16	27	90

The Catalan *anar* 'go' + Inf and the Spanish *ir* 'go' *(a)* + Inf were originally purposive constructions, which referred to a motion made with the aim of carrying out a certain act: 'go in order to do sth'. Some morphological characteristics of the medieval constructions are different from the properties of the modern ones.

The main difference is that in some medieval occurrences the verb 'go' and the infinitive are linked together with a preposition. In the Catalan text there are occurrences with the prepositions *a*, *per*, and *per a*, while in the Spanish texts there are variants used with the prepositions *a*, *por, pora*, and *para*. The Catalan prepositions *per* and *per a*, and the Spanish *por, pora*, and *para*, just like the preposition *a*, are purposive prepositions. However, I regard variants used with them as independent constructions and I do not include these occurrences in the study, because in them the verb *anar/ir* always conveys its literal meaning 'go'. In contrast, occurrences with the preposition *a* should not be left out of consideration, given that this is the version which grammaticalized into an immediate future tense in Spanish.

Occurrences in which the purposive preposition *a* 'to' appears between the finite verb 'go' and the infinitive are very rare at the beginning, but later their use spreads in Spanish texts and it grammaticalizes into a future tense. It is worth considering whether the use of the preposition played a role in the different evolutions of the Catalan and Spanish constructions. The analysis of occurrences in the corpus reveals that the prepositional version of the construction had a more restricted use in the medieval language state than the use of the construction without a preposition, and it tends to appear in later texts. Among the medieval Catalan chronicles, the prepositional variant only appears in Muntaner's Chronicle sporadically, which means a total of 6 occurrences, as against 234 occurrences without a preposition. Desclot's Chronicle contains two occurrences with the preposition *a* 'to' as shown in Table 3b, but both appear in a title, and we know that titles were often added later to historical texts, and, as a consequence, these should be considered as belonging to a later language stage (see Pérez Saldanya 1996: 74). In Spanish the situation is similar. In the *Cantar de mio Cid*, the text that reflects the earliest language state, there are only non-prepositional forms, but in later texts the prepositional variant extends its use to a greater degree than in Catalan (90 occurrences with a preposition and 297 without it, while Catalan texts contain 36 prepositional forms and 391 without a preposition). As for the early use of the preposition *a* in the Catalan construction, Segura-Llopes (2012: 122) states that it first appeared in texts written in the Valencian dialect, while it is not characteristic in the Eastern Catalan dialects at the beginning. Given this dialectal distribution and the fact that the neighbouring Aragonese and Castilian languages[29] show a frequent use of the preposition *a* in this construction, according to Segura-Llopes it is reasonable to suppose that the use of the preposition *a* spread thanks to the influence of these languages.

The prepositional variant of 'go' + Inf, both in Catalan and Spanish, relates to a use of the verb *anar/ir* 'go' in its literal meaning. It is worth mentioning results provided in Bruguera (1981: 38–39), who compared various manuscripts of the Chronicle of James I. He observed that certain occurrences appeared with a preposition in certain manuscripts, but without it in others. Bruguera (1981: 41–42) supposes that the prepositional version proliferated in Catalan over the course of time in order to differentiate those cases where the verb *anar* appears as a main verb meaning 'go'. According to this supposition, the preposition *a* appears in later texts in order to convey the motion meaning in places where earlier it was not needed (see examples provided in Bruguera 1981: 39). Bruguera's supposition seems to be correct on the basis of various observations. First, it is in line with the results presented above concerning the use of prepositional forms. Second, if the verb *anar* is used as a main verb with the motion meaning, the preposition *a* is obligatory in current Catalan too. Third, prepositional uses in the present tense version of the Catalan GO-construction, which supposedly reflects a more advanced stage of grammaticalization, are absent in narrative, past reference contexts. The sporadic prepositional instances with the verb 'go' conjugated in the present both in Spanish (five occurrences) and Catalan (four occurrences) are not examples of a historical present use, but 'real' present tense forms, with present tense reference, as can be observed in the Catalan example (18), which shows the construction in a general present tense description:[30]

(18)

*és molt vell, que no pot anar: no **va** sinó de matí **a dir missa** a una parròquia qui està prop d'allí a on posa*

'he is very old, he can hardly walk: he doesn't **go** anywhere but, in the mornings, to a vicarage near to his residence **in order to celebrate mass**'

(Ep2 178)

The situation in Spanish is slightly more complicated, since there are more prepositional occurrences and, furthermore, it is the prepositional version that grammaticalized into a future tense in current Spanish. The present book does not aim to investigate the evolution of the Spanish construction, but some considerations seem to be useful. Since the use of the non-prepositional variant is also documented in early Spanish texts and seems to convey a past meaning at least in certain contexts, it is possible that similarly to Catalan, in Spanish too it was the use of this grammaticalizing construction which motivated the addition of the preposition *a* in order to differentiate those cases where the verb *ir* appears as a main verb. According to this idea, although the process of grammaticalization of the non-prepositional version into a past marker failed in Spanish, the prepositional construction could be the starting point of a new grammaticalization process into a future marker. As for Catalan, Radatz (2003) claims that after the past meaning of the Catalan *anar* + Inf had been consolidated, the use of the preposition *a* became obligatory in the construction used in the literal meaning in order to formally differentiate both constructions. Thus, the new prepositional purposive construction (*anar a* + Inf) acquired the possibility of grammaticalizing into a future tense in Catalan as well, just as had already occurred in other Romance languages. Radatz (2003) supposes that in current Catalan there is an ongoing process of grammaticalization of *anar a* + Inf into an immediate future tense. If this is correct, in both Catalan and Spanish the same processes can be attested, only on a different time scale.

In summary, on the basis of the results of my study and those of other authors, it may be hypothesized that the original version of the construction under study was without any preposition in both languages. It may be supposed that the prepositional version proliferated in Catalan (and perhaps also in Spanish) in the course of time in order to differentiate those cases where the verb 'go' appears as a main verb. Therefore, when investigating the beginnings of semantic change, we have to take into account the non-prepositional occurrences.

2.5.3 Occurrences of the construction 'go' + Inf in the present vs. past tenses

After considering the use of prepositions in the construction, another morphological characteristic to be investigated is the tense of 'go'. The verb *anar/ir* 'go' can appear in medieval Spanish and Catalan texts conjugated in either past or present tense. The present tense version does not only occur in past contexts, but also in descriptions with present tense reference (see examples in Colon 1959/1978: 127–128). That is why Mendeloff (1968) supposes the coexistence of two *va* + Inf constructions: a perfective form and a formally analytic inchoative present tense. In the present research the first form in past reference contexts is relevant, because the construction is supposed to have been grammaticalized into a past tense in these narrative contexts (cf. Colon 1959/1978: 120; Pérez Saldanya 1996). Both the present and past tense versions of the Catalan GO-construction can occur in a preterit context. We have already seen in Section 2.3 that examples of the construction which are formally in the present tense, but are used in a past context, are regarded in some authors as instances of a historical present usage, while in others as grammaticalized forms. These two hypotheses have been related to two different scenarios of semantic change: (a) it is the present tense variant (interpreted as historical present) that undergoes semantic change, while the variant with the verb 'go' conjugated in the past is abandoned, or (b) semantic change begins with the variant conjugated in the past, while the present tense versions in a preterit context are later developments, proliferating after the past meaning of the construction has been consolidated.

As regards the distribution of occurrences according to the verbal tense of the verb 'go', the analysis of the corpus reveals that the present tense variant only occurs without a preposition, in a narrative context with a past reference. In the case of Catalan, we can take into consideration primarily data originating from the three famous medieval chronicles, since they contain most of the early occurrences of the construction under study. The Chronicle of James I, reflecting the most archaic language state, also provides two occurrences which have the verb *anar* in the present tense, while in Desclot's Chronicle almost a quarter of the occurrences are conjugated in the present. Finally, in Muntaner's Chronicle the proportion is different: there are more than twice as many occurrences in the present as in the past tense. From later texts we have only a few examples (cf. Tables 3a and 3b), but they are revealing: present tense variants are almost exclusive, and all of them can be regarded as grammaticalized forms (see occurrences in Section 2.6.3.2, especially (38), and considerations presented in 3.4.2.2). The analysis of occurrences in their contexts confirms that the present tense version of the Catalan GO-construction in a preterit context is a later development, which proliferated as a consequence of the ongoing grammaticalization of the construction, and the earliest occurrences had the verb *anar* 'go' in the perfective past tense.[31] On the basis of our corpus data, it can be argued that present tense occurrences do not reflect a historical present use, but can be interpreted as grammaticalized forms (see Juge 2006, 2008, and Section 2.6.8, which deals with the semantic change of the construction).

The Spanish texts of the corpus do not reflect the same spread of present tense occurrences over time. Moreover, occurrences conjugated in the present predominate in the two earliest texts, while the others show the preponderance of past tense occurrences. However, we should not conclude from this observation that we are dealing with the opposite process than the one occurring in Catalan. This distribution can, in fact, be attributed to the genre of texts. The two earliest texts analysed here are epic poems, usually characterized by a free use of tenses. This special alternation of different tenses concerns not only examples of the GO-periphrasis, but other verbal forms as well, which also appear conjugated in the present with a past reference. In order to compare the situations in Catalan and Spanish let us consider the following Spanish (19)–(20) and Catalan (21) text fragments:

(19)

Don <u>legan</u> los otros a minaya albarfanez *se*   *uan*   *homilar*
                                                    Refl go.3Pl.Pres bow.Inf
'When the others <u>arrive</u> they **go to bow** to Minaya Álvar Fáñez,

Quando <u>lego</u> auegaluon donta oio <u>ha</u>
when Avengalvón <u>arrived</u>, when he <u>is</u> in view,

Sonrrisando se dela boca **hyua-lo**   *abraçar*
                                  go.3Sg.ImpP-Clit   embrace.Inf
smiling from his mouth, he **was going to embrace** him,

Enel ombro lo <u>saluda</u> ca tal es su husaie
on the shoulder he <u>kisses</u> him, for such is his custom.'[32]

(Cid 1516–1519)

(20)

Luego alas manos la <u>prisieron</u> & dentro enla barqua la <u>metieron</u>
'Then they <u>took</u> her hands and <u>got</u> her into the boat

La barqua **van**   *Rjmar & luego sse meten ala mar*
                 go.3Pl.Pres pull.Inf
They **pulled/pull (= they go to pull)** the boat and then they <u>put out</u> (Pres) to sea

Luego <u>alçaron</u> las velas toda la noche <u>andan</u> alas estrellas
Then they <u>set</u> (PerfP) the sails and <u>go</u> all through the night by the light of the stars'

(Mar 69r)

(21)

E la host <u>eixí</u> en tal manera que <u>aconseguiren</u> los turcs e **van**   *ferir*   *en ells,*
                                                            go.3Pl.Pres attack.Inf
sí que en aquell jorn <u>mataren</u> ben mil hòmens de cavall, de turcs, e ben dos mília hòmens de peu.
'And the host <u>went out</u> in such manner that they <u>came upon</u> the Turks and **attacked (= go to attack)** them; so, on that day, they <u>killed</u> a thousand Turkish horsemen and full two thousand men afoot.'

(Munt II 78,43–44 – 79,1–2)

The fragment (19) taken from 'The Poem of the Cid' is characterized by the irregular alternation of different tenses. The past reference text contains a verbal form conjugated in the preterit (*lego* '(he) arrived'), but several present tense forms as well (*legan* '(they) arrive', *saluda* '(he) kisses', literally '(he) salutes'). Of the two instances of *ir (a)* + Inf, one shows the verb *ir* conjugated in the present (*uan* '(they) go'), while the other is in the continuous past tense (*hyua* '(he) was going'). In the fragment in (20) from 'The life of Saint Mary of Egypt' the occurrence of *ir (a)* + Inf is used together in the same context with other historical present forms (*sse meten* '(they) put out', *andan* '(they) go'), which alternate with past tense verbal forms (*prisieron* '(they) took', *metieron* '(they) got', *alçaron* '(they) set (PerfP)').

In contrast, the Catalan fragment in (21) shows a consequent use of preterit forms (*eixí* 'went out', *aconseguiren* '(they) came upon', *mataren* '(they) killed') in a past reference context, broken only by the present tense occurrence of the GO-periphrasis (*van ferir* '(they) go to attack/attacked').

The comparison of the use of tenses in the Spanish and Catalan texts of the corpus reveals that present forms in Spanish were not spread over time in the narrative genre, either. Among the narrative texts there is only one, the *Cuento de Tristan de Leonis*, that contains occurrences conjugated in the present, but used with a past tense reference. In contrast with the Catalan narrative texts, not only the *ir (a)* + Inf periphrasis is conjugated in the present tense in this text, but there are other present tense forms, which makes it likely that we are dealing with a historical present use. More than half the relevant contexts contain other present tense forms near to the occurrence of the *ir (a)* + Inf periphrasis. They are mostly motion verbs, like the verbal form *se lieuan* '(they) get up' in (22):

(22)

uan   se   ferir   de tan grand coraçon que los cauallos & los caualleros
go.3Pl.Pres Refl attack.Inf

fueron en tierra juntos en vno E despues <u>se lieuan</u> los caualleros & metieron mano alas espadas & fueron se el vno contra el otro

'they **go to attack/attacked** each other so resolutely that both the horses and the knights fell to the ground together. Then the knights <u>get up</u>, they grabbed their swords and then rushed at each other.'

(Tris 12v)

The other curiosity of the Tristan text is that it uses various motion verbs in the historical present (cf. *caualgan* '(they) ride/are riding' and *viene se* '(he) arrives' in (23)), and the verb *ir* frequently appears conjugated in the present tense also as a main verb, primarily in battle scenes (cf. *vase* '(he) goes' in Example (24)), but in other places as well. Similar phenomena cannot be observed in the Catalan texts.

(23)

*el torno le las saludes & **caualgan** todos & **viene se** ala corte con grand alegria & todos los dela corte aujan grand gozo dela tornada de tristan*

'he returned the greeting and they all **ride** and he **arrives** at the court very glad and everybody in the court was very pleased with Tristan's return.'

(Tris 10r)

(24)

*tristan tomo la espada enlas manos & **vase** contra el & dixo*
'Tristan took his sword in his hand and he **goes** up to him and said'

(Tris 16r)

In summary, as regards the verbal tense in which the verb 'go' is conjugated, the following results have been obtained. In Catalan, the present tense version of the GO-construction proliferated in later texts, i.e. it seems to be a later development in consequence of the ongoing grammaticalization of the construction. The Spanish texts of the corpus do not reflect the same spread of present tense occurrences. Supposedly, this would mean either that a similar grammaticalization process did not begin in this language or, if it did, it was interrupted. The contextual analysis has suggested that present tense occurrences of the GO-construction in Spanish texts, in contrast to Catalan, can be interpreted as historical present forms.

2.6 A new hypothesis on the formation of the Catalan GO-preterit

2.6.1 Introduction

In the process of meaning construction, lexical and contextual information interact to produce meaning. In order to reveal how semantic change in the motion construction under study may have occurred, we have to determine what meaning historical occurrences convey and, especially, what meaning the GO-verb conveys within them. Guessing meanings cannot rely exclusively on the researcher's subjective judgments. Some aspects of meaning attribution in historical research will be discussed in Section 3.5. In the present section I compare the use of 'go' + Inf with the use of other, functionally similar periphrases. This method is proposed in Fischer (2004; 2007: 15–17), who emphasizes the usefulness of comparison between formally or functionally similar constructions belonging to the same language state.

The current use of the Catalan *anar* + Inf construction as a perfective past tense is the result of an evolution that diverges from the widely attested tendency of motion verb constructions to evolve into future tenses in various languages around the world. The development of the Spanish *ir a* + Inf is a good example of this tendency. However, at first sight the Catalan and Spanish constructions seem to have similar roots. The similarity of contexts, functions, semantic nuances, and morphological characteristics of the Catalan *anar* 'go' + Inf and the Spanish *ir* 'go' *(a)* + Inf constructions in medieval texts suggest that they have a common origin and prompt us to compare both constructions in more detail. Such a comparison seems to be useful, because the evolution of the Spanish GO-construction fits one of the cross-linguistically frequent tendencies of semantic change in motion verb constructions. The formation of the Spanish immediate future tense from the GO-construction *ir* 'go' *(a)* + Inf is a typical representative of a well-documented process with various typological parallelisms. In contrast, the development of the Catalan construction has not followed the typical pathway of purposive constructions; moreover, it lacks well-documented typological analogies. A comparative study of the histories of both constructions can shed light on the motives for the divergence of the Catalan GO-construction. Information concerning cross-linguistically recurrent tendencies of meaning change can be used as data when we do not have enough evidence concerning a certain historical process for lack of written sources (cf. Heine 2003/2005). Accordingly, this type of study can also be used as a complementary data source in cases in which we have surviving historical texts from a language.

Relying on this similarity and the fact that the meaning 'intention' is an intermediary stage in the semantic evolution of purposive constructions towards futurity meanings in several languages of the world (cf. Bybee and Pagliuca 1987; Bybee et al. 1994; Bybee 2002: 181; Heine and Kuteva 2002: 161–163), among others in Spanish, I have assumed that this meaning in which the subject's intentions are emphasized could also have been present in Catalan. I have attempted to investigate whether the initial stage of semantic change in the Catalan construction followed the typical grammaticalization cline from motion verbs to intentionality markers observed in various languages around the world (cf. Bybee and Pagliuca 1987; Bybee et al. 1994; Bybee 2002: 181; Heine and Kuteva 2002: 161–163). In order to test this hypothesis, in addition to the Catalan and the Spanish GO-constructions, I have included medieval Catalan and Spanish occurrences of the periphrasis *pensar (de)* 'think of' + Inf in the analysis. This construction expresses the subject's intention to carry out the act described in the infinitive and was used in both medieval Catalan and Spanish. If its use in the language state under study turns out to be similar to that of the *anar* + Inf construction, this would be supporting evidence in favour of my hypothesis.

I have also considered the hypothesis provided in relevant literature according to which the Catalan construction under study was an inchoative periphrasis in the medieval language state (see Section 2.3.3). In order to test this hypothesis, I have included occurrences of the inchoative construction *començar (a/de)* 'begin to' + Inf in medieval Catalan and Spanish. If the inchoative hypothesis is correct, the uses of this inchoative construction and the GO-construction should be similar. I have attempted to investigate whether historical data really reflect such a similarity.

In summary, in addition to the Catalan and Spanish GO-constructions, I have included in the analysis medieval Catalan and Spanish occurrences of the periphrases *pensar (de)* 'think of' + Inf and *començar (a/de)* 'begin to' + Inf as well, because of the similarity of their contexts and functions in medieval texts (cf. Fischer 2007: 15). Such a comparison seems to be relevant, because the constructions formed with the verbs 'go', 'think', and 'begin' all appear in similar contexts and are frequently used in descriptions of battle scenes. In other words, in a first approach, both hypotheses seem to be correct: the one which assumes an inchoative meaning and the other which supposes that the Catalan *anar* + Inf was an intentional periphrasis. In order to decide between these rival hypotheses and to detect to what degree these three constructions are similar and what kind of similarity is involved, we need a deeper contextual analysis. The corpus of the study contains a total of 1,903 occurrences with the finite verb conjugated in the perfective past indicative,[33] namely, 814 occurrences of the medieval Spanish and Catalan GO + Inf periphrases, 221 examples of the THINK OF + Inf periphrases and 868 of the BEGIN TO + Inf constructions. The research has relied primarily on qualitative analysis of the occurrences in their broad contexts, although I have applied quantitative methods, too (cf. Navarro 2008).

In this section I aim to answer the following research questions, repeated here for convenience:

iv. What meaning did the Catalan *anar* + Inf construction express at intermediate stages of its grammaticalization?

v. What similarities and differences can be revealed in the uses of the Catalan *anar* + Inf and the Spanish *ir (a)* + Inf constructions on the one hand and other periphrases used in similar contexts on the other in the Middle Ages? What conclusions can be drawn from these similarities and differences in use?

2.6.2 The status of the verbs 'go', 'think', and 'begin' in the constructions under study

The description of the uses of these three periphrases leads to a terminological question. They are all periphrases in the sense that they express a unified notion by the combination of a finite verb and an infinitive. However, we should ask whether the finite verbs of these constructions are all auxiliaries or not. This problem is relevant from the point of view of the grammaticalization of the Catalan *anar* + Inf, since in this process a main verb meaning 'go' grammaticalizes into a preterit auxiliary. During semantic change, in some occurrences the verb *anar* is interpretable as a main verb, while in others it must already be considered an auxiliary. There are also ambiguous cases, where it would be hard to decide whether the verb *anar* conveys its original meaning or a new one. In sum, it is worth considering the status of the conjugated verbs in the constructions formed with 'go', 'think', and 'begin' before making the comparison.

Romance languages use several combinations of a finite verb and a second main verb in the infinitival, gerundive, or participial forms. Some of them are complex verb tenses, in which the finite component has a mere grammatical meaning (indicating number, person, tense, and mood), others express aspectual or modal meanings, and in yet others both verbal forms express independent notions.[34] The relevant literature uses several criteria to determine whether these constructions are periphrases or not, i.e. whether the finite element of such combinations is an auxiliary or not. However, neither of these criteria is valid in itself. A typical auxiliary does not have a full lexical meaning and carries only grammatical information. Desemanticization is sometimes regarded as the main criterion of auxiliarity. For instance, Gili Gaya (1961: 109, cited in Martínez Gómez 2004) claims that in the BEGIN TO-periphrasis in current Spanish, the finite verb is not an auxiliary, because its inchoative character is due to the meaning of the finite verb, and not to the complete verbal phrase. Others conceive of auxiliarity as a gradual notion and accept forms which do retain their lexical meaning as auxiliaries (Martínez Gómez 2004; cf. also Olbertz's (1998) terms: *lexical verb, semi-auxiliary*, and *auxiliary*).

Auxiliaries and finite verbs of aspectual and modal constructions have some characteristics in common; the two verbal elements, for instance, have a closer relationship in them in the sense that they express a unified notion, while in other constructions, where a main verb takes a verbal complement, verbal elements express independent notions. In Neves's (2000: 25–65, cited in Wachowicz 2007: 226–227) categorization modalizers, aspectual verbs and auxiliaries all belong to the category of verbs that do not predicate (cf. also Wachowicz 2007: 233). In this sense, the verbs discussed in the present volume may all belong to the same category.

The three cases of periphrasis discussed in the present book correspond to the three categories proposed by Martínez Gómez (2004), who makes a distinction between fully grammaticalized auxiliaries ('go' in some contexts, see Example (38)), partially grammaticalized or desemanticized[35] auxiliaries ('think', and 'go' in some contexts, cf. occurrences in (41)–(43) and (72)), and slightly desemanticized auxiliaries ('begin', cf. Section 2.6.3.4). However, the verb 'go' shows a heterogeneous behaviour usual in auxiliary verbs as pointed out by Wachowicz (2007: 226): '[I]n some contexts they recover the sense of full verbs: the process of grammaticalization may exhibit different stages in one and the same synchronic cut, whereas the same does not occur with aspectualizers'. This phenomenon can be observed in uses of the GO-periphrasis, which shows different degrees of grammaticalization in different contexts in the language state under study. The verb 'go' in some contexts retains its original lexical meaning (that is, we are not dealing with a periphrastic use in them), in some contexts it seems to express intentional modality, while in others it seems to be a 'true auxiliary' (i.e. the whole construction expresses a past tense value).[36] The BEGIN TO- and THINK OF-constructions seem to express aspectual and modal values. The verb 'begin' differs from 'go' in the sense that it retains its original lexical meaning in all contexts, while the verb THINK conveys the meaning 'intention' in periphrastic uses.

To sum up, in the relevant literature there is no criterion which would be commonly agreed to determine whether the finite element of a verbal construction is an auxiliary or not. Some criteria can be used in certain languages, but not in others, and even within the same language there are verbal elements unequivocally regarded as auxiliaries that do not satisfy all criteria (cf. Kenesei's (2000) research on auxiliaries in Hungarian). The periphrases discussed in the present study seem to be constructions possessing aspectual or modal values, although the verb 'go' already seems to be a true auxiliary in some contexts.

Because of the terminological problems discussed above, whenever it is possible, I avoid the use of the term *auxiliary* throughout the present volume when referring to the finite verbal element of the GO + Inf, THINK OF + Inf and BEGIN TO + Inf periphrases. However, because the topic of the present research is the evolution of the auxiliary of the Catalan perfective past tense from a motion verb, the question of when we may legitimately apply the term *auxiliary* is an essential one. In other words, from what point in time can we regard this verbal combination as a grammaticalized construction, and in which contexts is the verb *anar* used as a main verb, conveying its lexical meaning? In our historical corpus, the verb *anar* appears in some contexts as a main verb, while it seems to function as an auxiliary in others. The situation is even more difficult given that these uses can coexist in one and the same language state. Since the process of semantic change is always characterized by ambiguity, in some cases it is practically impossible to decide whether we are dealing with a main verb or an auxiliary use. In these cases, I have followed the methodological principle according to which we have to interpret the construction in the sense of its literal meaning whenever it is possible, in order to avoid projecting back upon earlier periods meanings accessible only in later ones. As presented below in Section 3.5, I have attempted to establish several formal criteria to decide whether the motion meaning is accessible in a given context. In contexts where the construction *anar* + Inf conveys a perfective past meaning, the verb *anar* may be regarded as an auxiliary. Since in complex verb tenses the finite verbal element is unanimously regarded as an auxiliary, in these cases I apply the term *auxiliary*, but in ambiguous cases I avoid it. In contrast, I use the expression *periphrasis* when referring to the constructions under study, since this term is widely used in grammars of Romance languages.

In the constructions *començar (a/de)* 'begin to' + Inf and *pensar (de)* 'think of' + Inf the status of the finite verb is uncertain in the sense that in the literature there is no useful criterion to decide whether it is an auxiliary (perhaps a semi-auxiliary) or a main verb. The THINK OF-periphrasis had an intentional meaning and the BEGIN TO-periphrasis an inchoative meaning in the language stage under study. For the purposes of the present research, it is not important to take sides on whether the finite verb in these periphrases is an auxiliary or a main verb, because I use them only for a semantic comparison. Hence, in the analysis of historical occurrences, I will use the term *periphrasis* or simply *finite verb*, which I hope is acceptable to everyone, given that using, for example, the term auxiliary in the case of the BEGIN TO-periphrasis, in which the finite verb retains its lexical meaning, would sound strange to some readers.

2.6.3 Entrenched local schemas in the uses of 'go' + Inf, 'think of' + Inf, and 'begin to' + Inf

2.6.3.1 Introduction

At the beginning, semantic change in grammaticalization is dependent on some specific contexts, while later the grammaticalizing construction appears in more contexts, which supposes an increase in the frequency of use (cf. Bybee 2003/2005). Early occurrences of the Catalan and Spanish 'go' + Inf constructions are attested – almost without exception – in certain types of contexts (emotionally charged scenes, principally descriptions of battle scenes). Not only are contexts of use highly constrained at this early stage, but the construction is preferred with certain infinitives. The relevant literature also mentions some frequently used phrases (e.g. *anar ferir* 'go to attack'), and the study of our historical corpora confirms the impression that several highly entrenched schemas predominate in the early use of these motion verb constructions: the Catalan verb *anar* and the Spanish verb *ir* are preferably combined with certain infinitives. As Colon (1959/1978: 130) points out, the infinitives in early combinations are all perfectives.[37] These highly entrenched collocations can serve as a starting point when analysing incipient stages of the grammaticalization of *anar* into a past tense auxiliary, because these contexts may have been the triggering environments for semantic change.

These phenomena are described in Bybee (2003/2005: 604) relying on the notions of token frequency and type frequency. She defines token frequency as 'the frequency of occurrence of a unit, usually a word or morpheme, in running text', while 'type frequency refers to the dictionary frequency of a particular pattern' (Bybee 2003/2005: 604). In another place, Bybee (2003/2005: 605) notes that 'the notion of type frequency can also be applied to grammaticizing[38] constructions by counting the different lexical items with which a construction can be used' (e.g. types of subject, infinitives etc. with which a certain item can be used). While high token frequency corresponds to a very strong or highly entrenched local schema and affects the nature of cognitive representation, high type frequency corresponds to the generality of schema. Type frequency and token frequency are closely related, since the increase in the number of lexical units which can co-occur with a certain grammaticalizing linguistic item triggers an increase in token frequency, which may induce different formal and functional changes (Bybee 2003/2005: 605). Frequency counting in the senses described above can provide useful information about the process of grammaticalization, since higher type frequency corresponds to a higher degree of grammaticalization (Bybee 2003/2005: 612). Bybee (2003/2005: 605–614) applies these frequency counting methods in a study concerning the process by which the English verb *can* has become an auxiliary, taking into consideration both token and type frequencies on a corpus of Old and Middle English texts. She makes a distinction between verbs of intellectual states or activities, verbs of communication, and verbs of skills, since these are the verb types which appear typically with *can*. After considering 300 occurrences of the construction *can* + Inf, she has the following results from the Middle English corpus concerning the different semantic groups of infinitives which can appear together with *can*:

1.   verbs of intellectual states or activities: 52 tokens, 18 types

    high token frequency: *see* 12, *deem* 6, *understand* 6, *espy* (discover) 5

    type frequency: 18 distinct verbs

2.   verbs of communication: 102 tokens, 31 types

    high token frequency: *tell* 30, *say* 29, *devyce* 'describe' 8

    type frequency: 31 distinct verbs

3.   verbs of skills ('know how to'): 26 tokens, 18 types

(Bybee 2003/2005: 611–612)

Bybee compares these Middle English proportions with the corresponding proportions in Old English. The comparison shows that although infinitives belonging to these groups could be used with *can* in Old English too, in Middle English more infinitives are documented in each group, and moreover, the token frequency of some combinations in the first two groups of infinitive types increases. At the same time, new infinitives appear with *can*. According to Bybee, both token frequency and type frequency are relevant in grammaticalization research, since both contribute to semantic bleaching. Following Bybee's method, I have examined the frequent collocations in the use of the Catalan *anar* + Inf and the Spanish *ir (a)* + Inf constructions, which may have triggered grammaticalization. I present the infinitive types entering the periphrases by describing entrenched local schemas detected in the historical texts.

The verb 'go' is combined with the following infinitives with a relatively high token frequency in the Catalan texts of the corpus:[39] *ferir* 'to attack, to hit, to wound' (77), *assetjar* 'to besiege' (20 + 3 synonyms), *besar* 'to kiss' (16), *pendre* 'to take, to grab' (17), *dir* 'to say, to tell' (12), *abraçar* 'to embrace' (10), *donar (colp)* 'to give (a blow)' (9), *brocar* 'to spur, to gallop' (9), *veer/veer-se (ab)* 'to meet sb' (12). In the Spanish texts the following infinitives are the most frequent: *ferir* 'to attack, to hit, to wound' (72), *cercar* 'to beleaguer' (47), *posar* 'to lodge, to accommodate' (27), *ver/verse (con)* 'to meet sb' (21), *abraçar* 'to embrace' (15), *besar* 'to kiss (13), *correr* 'to prowl' (10), *fablar* 'to speak, to talk' (10), *reçebir* 'to receive, to welcome' (7). The similarity of uses suggests that phrases of the type 'go' + Inf in several neighbouring languages probably have common roots.

2.6.3.2 Anar *'go'* + Inf / Ir *'go'* (a) + Inf

a. Anar ferir/ir (a) ferir[40] *'go to attack/to hit/to wound'*

In the corpus of the present research the verb *anar* was combined most frequently with the infinitive *ferir*. There are so many occurrences of this collocation that it seems to be a deeply entrenched schema: from 391 periphrases without a preposition 77 are instances of *anar ferir*, i.e. almost every fifth occurrence of *anar* + Inf contains the infinitive *ferir*. The type *ir (a) ferir* is also the most frequent combination in the Spanish texts of the corpus, in which 72 of the 297 occurrences appear with this infinitive. The periphrasis 'go' + *ferir* is described as a very frequent phrase in Old French, as well (Gougenheim 1929/1971; Colon 1976/1978). The Catalan occurrences are taken from the chronicles of James I, Desclot, and Muntaner, and their distribution is also noteworthy concerning the tense of the finite verb.

Table 5. Distribution of the construction *anar ferir* according to the tense of *anar*

	Jau	Desc	Munt
Preterit	6	18	4
Present/preterit syncretism	1	-	2
Present	1	5	40
Total	8 (~17.77%)	23 (~23.95%)	46 (~19.65%)
Total *anar* + Inf	45	96	234

Table 5 shows that, while the finite verb of the construction (*anar*) is conjugated mostly in the preterit in earlier texts or it is a 1Pl syncretic form, which is also interpretable as a preterit, later on a spread of the present tense variant can be observed. Present tense variants already predominate in Muntaner's Chronicle. According to the historical present hypothesis, these proportions would reflect the spread of the use of the historical present. I argue that examples in the present tense should rather be considered grammaticalized forms, because if we were dealing with a historical present usage, it should concern other verbal forms too, and not only occurrences of the GO-periphrasis. This distribution confirms the fact that the reanalysis of the construction and its semantic change begins in these frequently used collocations.

In contrast, the verb *ir* is conjugated in the preterit in most of the occurrences of the parallel Spanish construction *ir (a) ferir*. The *Cantar de mio Cid* contains two exceptions, which are probably the historical present, as attested by the tense of neighbouring verbal forms:

(25)

La seña <u>sacan</u> fuera de valençia <u>dieron salto</u>
'The standard they <u>bring out</u>, from Valencia they <u>sallied</u>,

Quatro mill menos xxx con myo çid <u>van</u> a cabo
four thousand minus thirty <u>go</u> together with my Cid

Alos çinquaenta mill **van**   *los ferir*   *de grado*
                     go.3Pl.Pres Clit strike.Inf
towards the fifty thousand they **strike them (= go to strike them)** gladly,'

(Cid 1716–1718)

Even more notable are those seven present tense occurrences which appear in the *Cuento de Tristan de Leonis*. Four of them co-occur with other present tense verbal forms in the same context, as illustrated above with Examples (22)–(24), also taken from the *Cuento de Tristan de Leonis*. There are also sporadic occurrences which are similar to those from the Catalan chronicles: in these cases, the only present forms in a preterit context are instances of *ir (a)* + Inf, and the infinitive describes an act which can be considered as the culmination of the battle scene. Let us consider the following Spanish example.

(26)

*& **van-se**   ferir   asi fuerte mente el vno contra el otro de tan*
    go.3Pl.Pres-Refl attack.Inf
grandes golpes que los cauallos & los caualleros fueron atiera todos juntos
'they **attack/attacked (= go to attack) each other** so hard and with such great blows that both the horses and the knights fell to the ground together'

(Tris 16r)

These present tense forms perhaps suggest that the grammaticalization process attested in Catalan may have begun in Spanish as well. However, in contrast to Catalan, Spanish narrative texts of the corpus do not reveal a later spread of the present tense variant of *ir (a)* + Inf.

It is worth examining this frequent combination from the point of view of contexts, as well. Examples of the construction 'go' + Inf can appear in various types of contexts. The most frequent context type is where the clause which follows the one containing the periphrasis refers either to the consequences of the action described in the infinitive of the GO-periphrasis or to the battle which developed after the act of attacking described in the infinitive, and its consequences. The description of the consequences of the battle is missing in some cases. In Catalan texts of the corpus there is only one occurrence which appears in a context where what is implied is not the carrying out, but the failure of the act of attacking. In Spanish there are a few sporadic examples in imperative contexts, too. The distribution of these context types is shown in Table 6.

Table 6. Distribution of context types of the constructions Catalan *anar ferir* and Spanish *ir (a) ferir*

		Catalan *anar ferir*			**Spanish** *ir (a) ferir*				
		Jau	Desc	Munt	Cid	GrCo	Tris	CrPC	
1.	(Description +) consequence	5	21	40	-	39	16	12	133
2.	Description	2	2	6	1	-	-	-	11
3.	Imperative	-	-	-	-	-	-	2	2
4.	Plan	1	-	-	1	1	-	-	3
	Total	8	23	46	2	40	16	14	149

I will illustrate each context type with an example.

1a Consequence

The most frequently attested context type is where the occurrence of the GO-construction is followed by the description of the consequences of the attack in the subsequent clause. The description of the consequences implies that the act of attacking was effectively performed. The Catalan (27) and Spanish (28) examples instantiate this typical use of GO + Inf with the dynamic verb *ferir* 'attack'.

(27)

E quant fo defora, puyí lo cavayl dels esperons e **anà** **ferir** *entre·ls*

go.3Sg.PerfP attack.Inf

sarraÿns, sí que·l primer colp n'abaté, ab lo pits del caval seu, ·IIII· a terra.

'And when he was outside, he dug his spurs into his horse and **went to attack/attacked** the Saracens, so that at first he threw four of them to the ground with the breast of his horse.'

(Desc III 93,23–26)

(28)

*E el torno contra el & **fue**-le   ferir   que le passo el escudo &*
                 go.3Sg.PerfP-Clit   attack.Inf

la loriga & metiol la lança por la carne

'And he turned towards him and **went to attack/attacked** him (in such a way) that he stabbed his shield and his coat of mail through and pierced him through with his lance.'

(Tris 11r)

In Examples (27) and (28) the subsequent utterances describe consequences of the action described in the infinitive. In this context the GO + Inf construction presents the event in question in its entirety, as a completed action, which takes place as a whole. The act of attacking described in (27) caused the death of four Saracens and that in (28) had the consequence that the enemy's shield and coat of mail were pierced, and he was pierced through with the lance. Although Examples (27) and (28) allow the motion-then-action reading too, the GO-construction seems to direct the attention to the very event described in the infinitive, which is presented in the context as an act charged with consequences. The contexts show that it does not only express motion, but, in its more grammaticalized versions, also an emotional charge and dynamism, inasmuch as it appears in the most tense and interesting moments of the story. Contexts similar to those provided in (27) and (28) may have triggered the grammaticalization process, since the content of motion is relegated to the background in them, the act described in the infinitive seems to be emphasized and the perfectivity of this act is implied.

1b Description + consequence

The second context type is very similar to the first one, the only difference being that the description of consequences does not directly follow the clause containing the occurrence of 'go' + *ferir*, but a description of the battle scene is inserted. However, the description of the battle itself and its consequences implies the actual performance of the attack in this context, too. Observe Example in (29).

(29)

*Barons – dix él –, los sarraÿns són grans gens, mas Déus serà ab nós; e cascú aja ferm cor e bona esperança en Déu, e firam en éls ardidament. – A aquest conseyl s'acordaren tots e **anaren ferir** e·ls sarraÿns, cavalers e sirvents. La batala fo molt gran, mas los sarraÿns no·u pogren durar e comensaren de fúger devés les muntanyes que hi eren grans; e·ls crestians, firén e talan, encalsaren-los tro en la montanya, que no pogren anar pus avant per les montayes qui eren altes e·ls boscs grans. En aquesta batala moriren bé ·II· mília sarraÿns, menys d'aquels que retengren a presó.*

'Lords – he said –, there are a great many Saracens, but God will be with us; you all shall take heart and have good trust in God and let us attack them vigorously. And they all agreed on this advice and **went to attack/attacked** the Saracens, knights and footsoldiers. The battle was very hard, but the Saracens couldn't resist any longer and began to escape towards the mountains, which in that region were high; and the Christians followed them, attacking them and stabbing them, to the mountains, because they could not go any further, because the mountains were high and the forest dense. In this battle about two thousand Saracens were killed, except for those who were taken prisoner.'

(Desc II 98,22–99,5)

2 Description

There are fewer occurrences used in a context where the narrator does not describe the consequences of attacking in detail, but only mentions in general that a battle developed. In a broader sense this can be interpreted as a consequence of attacking and implies that the attack was actually carried into effect, as in (30):

(30)

*encontraren-se ab la host dels sarraÿns e **anaren ferir** en éls; sí que la batayla fo gran e forts.*

'they met the troops of the Saracens and **went to attack/attacked** them; and so the battle was strong and hard.'

(Desc II 103,15)

3 Imperative

Among Spanish examples of 'go' + *ferir* there are two which appear in an imperative context. These present tense forms do not appear in a narrative context, thus are not relevant in the present research.

4 Failure of the act, plan

In the Catalan corpus there is just one context where it can be inferred that although the attack was begun, it failed: the aim of the motion described by *anar* was not achieved.

(31)

*e de l'altra part estaven los sarraïns ab les adargues e espaes treytes de la .I.ᵃ e de la altra part e no·s gosaven escometre. E, cant entraren los cavalers ab los cavals garnits, **anaren-los ferir**. E era tanta la multitut de la gent dels sarraïns, que·ls pararen les lances, e els cavals dreçaren-se per ço car no podien passar per la espessea de les lançes, sí que agren a fer la volta.*

'On the other side were the Saracens with their bucklers,[41] and swords drawn on a line, but neither dared to close. And when the knights with their armoured horses came in, they **went to charge** them; but so great was the multitude of the latter, that their lances stopped them, and the horses reared up as they could not get through the thick ranks of the lances. So they had to turn.'

(Jau 85)

The occurrence in (31) originates from the Chronicle reflecting the most archaic language state, which perhaps refers to the possible existence of uses where the periphrasis may have referred to the beginning of the act described in the infinitive. However, this hypothesis is difficult to confirm, because we do not have enough written sources from this early language state. Moreover, the example in (31) allows for another reading, according to which the attack was performed in a certain sense. In utterance (31) the clitic *los* refers to *los sarraïns* 'the Saracens'; they are identified as the direct object of the infinitive *ferir*, in other words, the aim of the attack was to charge the Saracens. However, the attackers could not reach them, because they could not break through the array of lances, and in this sense the attack failed. At the same time, we can interpret the fragment in (31) that they performed a certain sort of attack, since they rushed at the enemy and they reached the lances. It is only a matter of interpretation whether we understand this event as a failed attack or as a fully performed attack.

Finally, it is worth noting that in Spanish there is a further context type where the construction conveys the meaning of a future plan. The use of the construction in this type of context may have led to the emergence of the futurity meaning.

b. Anar pendre *'go to take'*

The infinitive *pendre* appears in the corpus conveying two different meanings: the 2 preterit occurrences mean 'to take (a city)', while the other 15 occurrences convey the meaning 'to take, to grab, to take in hand', which is a punctual action.

Table 7. Distribution of occurrences of *anar pendre*[42]

	Jau	Desc	Munt
Preterit	2	2	1
Present/preterit syncretism	3	-	-
Present	1	1	7
Total	6	3	8

The tense of the finite verb, similarly to the case of 'go' + *ferir*, reveals the ongoing process of grammaticalization. A similar construction is used also in Spanish: *ir (a) prender* 'go to take'.

c. Anar besar *'go to kiss'*

The infinitive *besar* is used in different constructions in the Catalan texts of the corpus, with a total of 16 occurrences: *anar besar (la mà/en la boca/lo peu/-se)* 'go to kiss (the hand/the mouth/the feet of sb/each other)'. Occurrences of the construction *anar besar* do not reveal the spread of present tense variants over time, because all of them appear in the same text, namely, in Muntaner's chronicle. Nevertheless, we can observe that in a quarter of the examples the finite verb is conjugated in the present, that is, they seem to be grammaticalized forms. The similar Spanish construction *ir (a) besar (la(s) mano(s))* 'go to kiss (the hand(s) of sb)', which occurs 13 times, mostly is conjugated in the preterit, except in the 'Poem of the Cid', where there are some present tense occurrences too, due to its genre.

d. Anar abraçar *(-se) 'go to embrace (each other)'*

In the case of the construction *anar abraçar* 'go to embrace' present tense variants appear in a later text, similarly to the constructions presented above.

Table 8. Distribution of occurrences of *anar abraçar*[43]

	Jau	Desc	Munt
Preterit	2	1	4
Present	-	-	3
Total	2	1	7

The similar Spanish construction *ir (a) abraçar* 'go to embrace' is frequent too, appearing 15 times.

e. Anar donar *(colp) 'go to give (a blow/hit)'*

Table 9. Distribution of occurrences of *anar donar (colp)*

	Jau	Desc	Munt
Preterit	-	1	-
Present	-	1	7
Total	-	2	7

The infinitive *donar* appears in several constructions in the corpus, with a total of nine examples: *donar .I. [44] gran colp* 'to give a great blow', *donar tal colp* 'to give such a blow', *donar tal* 'to hit so'. It occurs almost exclusively with the present tense conjugation which shows that it is not about a real motion, but that the verb *anar* conveys a different meaning in this phrase. The corresponding construction in Spanish, *dar golpe* 'to give a blow' (two examples), conveys the same meaning.

f. Anar assetjar/tenir setge *'go to besiege'*

Table 10. Distribution of occurrences of *anar assetjar/tenir setge*

	Jau	Desc	Munt
Preterit	1	8	9
Present/preterit syncretism	5	-	-
Present	-	-	-
Total	6	8	9

The finite verb of the phrase *anar assetjar*[45] 'go to besiege' is either conjugated in the preterit, or it appears in the syncretic 1Pl form. The five 1Pl occurrences can be interpreted as conveying a past reference, as suggested by the tense of surrounding verbal forms. However, the original motion meaning of *anar* can be attributed in each example, since in order to perform the action described in the infinitive (to besiege) a dislocation in space is needed. The new location is the city to be besieged. Consequently, the combination of *anar* with this infinitive cannot be regarded as a periphrasis: the use of the verb *anar* is necessary in the narration to describe the change of scene. Thus, the case of the infinitive *assetjar* is different from that of *ferir*. The aim of the motion, i.e. the new place, appears in the utterance as the direct object of the infinitive, as in Example (32):

(32)

anà assetgar Leyda
'(he) **went to besiege** Leyda'[46]

(Desc II 8,2)

The phrases *anar combatre* 'go to combat/to assault (a city)' (5 examples) and *anar barrejar* 'go to raid (a city)' (3 examples) behave in a similar way. The meaning of the Spanish phrase *ir (a) cercar* 'go to besiege' (47 occurrences) is equivalent to the Catalan *anar assetjar*. Similarly, the 10 instances of the construction *ir (a) correr* 'go to prowl' are members of an entrenched schema.

These and similar examples suggest that the present tense variant of the GO-construction cannot be simply interpreted as a historical present use, because if it was, it could be used with any kind of infinitive. However, the infinitive *assetjar*, which allows for a literal reading of *anar*, does not occur with present tense *anar*. Finally, it is worth noting that these contexts also allow for an inference to the effect that the act described in the infinitive was actually performed.

g. Anar veer[47] *'go to see/to visit',* anar veer-se[48] (ab) *'go to meet (sb)'*

The collocations *anar veer* 'go to see/to visit' (seven examples) and *anar veer-se (ab)* 'go to meet (sb)' (three examples) also allow for the literal motion reading.

The former appears in Muntaner's Chronicle twice with the preposition *a*, which shows that the verb *anar* conveys its original motion meaning.

Table 11. Distribution of occurrences of *anar veer* and *anar veer-se (ab)*

	Jau	Desc	Munt
Preterit	-	1	4
Present/preterit syncretism	5	-	-
Present	-	-	-
Total	5	1	4

The phrases *anar veer* and *anar veer-se (ab)* are not used with a present conjugation. All occurrences convey the motion meaning, at least this reading results in a meaningful interpretation, since the person to be visited is at a different place. Sometimes an adverb of place expresses this location, as in (33):

(33)

*E con ja us he dit, lo senyor rei d'Aragon **anà** una vegada **veure** lo papa en Roma,*

'And as I have told you already, the Lord King of Aragon **went**, on one occasion, **to see** the Pope in Rome,'

(Munt II 48,12–14)

The construction *anar visitar* 'go to see/to visit' is synonymous. It is attested twice in Muntaner's Chronicle, both with a preterit *anar*. It also has a prepositional version in the same text, which suggests that the addition of the preposition *a* in order to differentiate those cases where the verb 'go' appears as a main verb may have already begun in the 14th century. Later occurrences of *anar visitar* all appear with the preposition *a*. Let us compare the following two examples:

(34)

*E con açò fo fet, **anà visitar** Saragossa*

'And when this was done he **went to visit** Saragossa'

(Munt I 171,3)

(35)

*lo senyor rei **anà** a Saragossa, **a visitar** la ciutat e **veure** madona la reina e els infants.*

'the Lord King **went** to Saragossa, **to visit** the city and **to see** my Lady the Queen and the Infantes'

(Munt I 135,33–35)

In Example (34) the goal of the motion event encoded by the verb *anar* is expressed as the direct object of the infinitive, while in (35) it is expressed with a prepositional phrase. The difference between these two occurrences originating from the same text suggests that the preposition may initially have appeared when the goal of the motion was described by a prepositional phrase which separated the finite verb and the infinitival complement.[49]

The semantically equivalent Spanish construction '*ir (a) ver*[50]*/verse (con)*' 'go to see/to visit/go to meet (sb)' is another entrenched schema with 21 examples among which there are also 8 prepositional occurrences. The prepositional variant appears in Spanish already in early texts. The verb *ir* is conjugated in the preterit; nevertheless, it seems to convey its literal meaning in the present tense variant, as well. Good examples can be found in the *Cantar de mio Cid* (cf. 36), which is characterized by a free alternation of tenses, due to its genre. In contrast to the present tense variants of the GO-construction in Catalan texts, in the *Cantar* we are dealing with an actual historical present usage, as confirmed by the morphology of surrounding verbal forms in the historical present (*adelinan, fincan*):

(36)

<u>Adelinan</u> *a posar pora folgar essa noch*

'They <u>head</u> for their lodging to rest that night,

Minaya **va**   **uer**   *sus primas do <u>son</u>*

          go.3Sg.Pres see.Inf

Minaya **goes to see** his cousins where they <u>are</u>,

Enel <u>fincan</u> los oios don eluira e doña sol

doña Elvira and doña Sol <u>set</u> (Pres) their eyes on him'

(Cid 2857–2859)

h. Anar dir/anar a dir **'go to say/tell'**

The Catalan construction *anar dir* appears 12 times in the corpus, the latest three occurrences being prepositional.

Table 12. Distribution of occurrences of *anar dir/anar a dir*

	Desc	*Munt*	*Per*	*Ger*
Preterit	2	-	-	3
Present	-	1	6	-
Total	2	1	6	3

The distribution of occurrences of the collocation *anar dir* well illustrates the grammaticalization scenario of *anar* + Inf. Desclot uses it twice, with *anar* in the preterit: in both contexts the person to be talked to is at a different place, that is, motion is necessary. Observe Example (37):

(37)

*E los misatgers del rey d'Aragó **anaren dir** la misatgeria al senescal de Tolosa*

'and the messengers of the King of Aragon **went to tell** the message to the seneschal of Toulouse'

(Desc V 145,9–11)

Muntaner also uses a present tense variant, which is probably a grammaticalized form, and later in the text *Perellós* we have another six occurrences, also with other infinitives of communication (*responder* 'to answer', *demanar* 'to ask'), with *anar* conjugated in the present. Observe the fragment in (38):

(38)

*E quan ells haguen ab mi parlat així llongament, me menaren en una gran muntanya, e digueren-me que regardàs en aut e en avant, vers lo cel, e io hi **vaig regardar**, e **van-me demanar** de quina color era ni cum me semblava ni ont era io. E io lus **vaig respondre** que a mi me semblava color d'or e d'argent de la fornal ixit. E llavores ells **me van dir**:*

'And when they had spoken thus long with me, they led me to a great mountain and told me to look up and forward towards the sky. And I **looked** there and they **asked** me what colour it was and what it seemed to me and where I was. And I **answered** them that to me it seemed the colour of gold and silver come out of the furnace. And then they **said to me**:'

(Per 49)

The text fragment in (38) describes a dialogue. The motion reading is excluded, because it would result in a nonsensical interpretation. If we assumed that present tense forms of *anar* in this text are historical present uses, we would have the following incomprehensible description, where the interlocutors are moving all over the place during the communication event:

??'And when they had spoken thus long with me, they led me to a great mountain and told me to look up and forward towards the sky. And I **?go (?to the sky) to look** there and they **?go (?to me) to ask** me what colour it was and what it seemed to me and where I was. And I **?go (?to them) to answer** them that to me it seemed the colour of gold and silver come out of the furnace. And then they **?go (?to me) to say to me**:'

The translation above, beyond the fact that the frequent changes of scene in it are very confusing, contains two phrases which are in contradiction with the deictic nature of the verb *anar* 'go', which encodes motion directed away from the speaker's location (cf. *van-me demanar* literally: ?'(they) go (to me) to ask', *me van dir* literally: ?'(they) go (to me) to say to me').

Finally, in the latest text (*Germanies*) the construction under consideration appears again with the verb *anar* conjugated in the preterit: however, the utterances containing them also contain an adverb of place which shows that they refer to a real motion in space, which is also confirmed by the use of the preposition *a* in all three cases:

(39)

*Y la major part de la gent dels que **anaren** a sa casa **a dir-li** que ixqués ...*
'and the greater part of those who **went** to his house **to tell** him to come out ...'

(Ger 288)

In Spanish, too, several infinitives of communication are combined with the verb *ir* 'go', among others *fablar* 'to speak/talk', *dezir* 'to say/tell' and *llamar* 'to call'.

In summary, occurrences of *anar dir* 'go to say/tell' reveal the grammaticalization process of '*anar* + Inf': (a) the incipient stage when, although the construction is used without a preposition, the verb *anar* conveys its original motion meaning, (b) a more advanced stage of grammaticalization when the verb *anar* functions as an auxiliary, and finally, as a consequence, (c) the preposition *a* appears in order to indicate that the verb *anar* is intended to convey its literal meaning.

After presenting the use of the *anar* + Inf periphrasis, in the following we will discuss the use of the intentional construction *pensar (de)* 'think of' + Inf.

2.6.3.3 Pensar (de) *'think of'* + Inf

The construction *pensar (de)* 'think of' + Inf was used both in medieval Catalan and Spanish to convey an intentional meaning. It can occur in both languages without preposition or with the preposition *de*, but in Catalan also the preposition *a* appears sporadically between the two verbal elements.[51] Since I have not detected any difference in meaning between these morphological versions, I regard them as stylistic variants. In Catalan, occurrences with the preposition *de* are the most frequent, while in Spanish the proportion of the variants without the preposition and with the preposition *de* is similar. The distribution of occurrences of *pensar (de)* 'think of' + Inf in the Catalan and Spanish texts are presented in Tables 13 and 14.

Table 13. Distribution of occurrences of *pensar (de)* 'think of' + Inf in Catalan texts of the corpus

	Jau	*Desc*	*Munt*	*Per*	
pensar de + Inf	15	14	147	-	176
pensar + Inf	-	-	4	2	6
pensar a + Inf	-	1	1	-	2
pensar (mixed) + Inf	-	2	11	-	13
Total	15	17	163	2	197

Table 14. Distribution of occurrences of *pensar (de)* 'think of' + Inf in Spanish texts of the corpus

	Cid	Tres	GrCo	Alf	Tris	Arn	Pier	CrPC	
pensar de + Inf	6	1	2	-	2	-	-	2	13
pensar + Inf	-	-	-	1	2	6	2	-	11
total	6	1	2	1	4	6	2	2	24

In Catalan texts there are also instances of the construction with two or more coordinated infinitives. In most of these cases the same preposition is used to introduce the infinitives. However, there are exceptions, regarded as stylistic variants, as for instance in Example (40), where the first infinitive is introduced by the preposition *de*, while the second by *a*. In other cases, only the first infinitive is introduced by a preposition, either *a* or *de*.

(40)

E pensaren   molt vigorosament de   combatre,   e nós a   defendre.
  think.3Pl.PerfP   Prep combat.Inf   Prep defend.Inf

'And they **thought of combating** (= **combated**) very vigorously and we **of defending** (= **defended**) ourselves.'

(Munt II 101,27–28)

Occurrences of the THINK OF-construction show a varied picture. Since it is an intentional periphrasis, a wide range of infinitives can be used within it. The subject can intend to perform a certain kind of motion (e.g. the Catalan *anar* 'to go' and *exir* 'to go out, to exit', or the Spanish *cavalgar* 'to ride' and *ir* 'to go') or a communication act (e.g. the Catalan *saludar* 'to welcome, to salute', or the Spanish *responder* 'to answer'), some punctual action (e.g. the Catalan *ferir* 'to attack, to hit, to wound') or a prolonged action or event (e.g. the Catalan *armar/armar (se)* 'to arm', or the Spanish *conplir (mandado)* 'to fulfil (a command)'). Given that I aim to use data gained from research on this construction in the investigation of *anar* + Inf, it is worth considering combinations with those infinitives that appear with the verb *anar* as well, and try to reveal similarities and differences in their uses. Motion verbs enter the periphrasis formed with *pensar* most frequently. This type of infinitive appears in more than a quarter of the Catalan examples, while this proportion is 10% in Spanish, but other dynamic verbs also enter the construction. Since Spanish texts of the corpus contain only a total of 40 examples of the construction, it is not surprising that there are not frequently used infinitives. Catalan texts contain more examples, thus typical collocations can be observed occurring with them. Infinitives expressing some kind of motion include, among others: *anar/anar (se'n)*' 'to go, to leave' (21), *fugir* 'to flee, to get away' (9), *venir/venir (-se'n)* 'to come, to arrive' (8), *entrar/entrar (se'n)* 'to go in, to enter' (8). The infinitive *ferir* 'to attack, to hit, to wound', which forms a typical phrase with the verb GO, can enter also the THINK OF-periphrasis: this combination occurs eight times in the Catalan texts.

The following examples show occurrences of *pensar* 'think' with infinitives describing momentary actions in medieval Catalan (41)–(42) and Spanish (43).

(41)

*E **pensaren** de **fugir** ab tota lur roba; sí que quant la*
think.3Pl.PerfP Prep get-away.Inf

host del rey hy fo junta, trobaren la vila desemparada.

'And they **got away** (= **thought of getting away**) with all the booty; so that when the king's troops arrived there, they found the city abandoned.'

(Desc III 69,16–18)

(42)

*e los turcs, qui veeren que els muntaven darrera, **pensaren** ab les*
                                      think.3Pl.PerfP

*sagetes **de** trer, e per desastre una sageta va ferir En*
         Prep take.out.Inf

Corberan, qui s'hac desarmat lo cap per calor, per lo pols. E aquí ell morí.

'But the Turks, seeing them come up after them, **took out** (= **thought of taking out**) their arrows. And unfortunately an arrow hit En Corberan, who had taken off his iron cap because of the heat and dust, and there he died.'

(Munt II 77,2–5)

(43)

*Tres dias e dos noches **penssaron** de andar*
                      think.3Pl.PerfP Prep walk.Inf

Alcançaron a myo çid en teuar e el pinar

'Three days and two nights they **went** (= **thought of going**), they caught up with my Cid in the pine wood of Tevar'

(Cid 970–971)

According to the literal meaning of *pensar (de)* 'think of', this verbal construction literally means that the subject has thought about carrying out an act, in other words, he has the intention of doing something. However, on the basis of the narrow context we are not able to determine exactly what semantic nuance the construction represents. In any case, we can say that the types of infinitives which enter the construction and the types of contexts in which it appears (battle scenes, emotionally charged scenes) show a similarity with uses of the GO-construction.

Examples (41)–(43) reveal that occurrences of *pensar (de)* 'think of' + Inf, with the finite verb in perfective past tense refer to a completed, perfective event. By using this construction, the speaker describes the event as connected with a previous intention of the subject, and one which is worth attracting the listener's attention. Due to the fact that the intention is expressed in a perfective tense, an inference arises to the effect that the action described in the infinitive was a fully performed action, that is, one that was actually carried out. This inference is supported by information from the broad context. All three examples are similar to uses of the *ir (a)/anar* 'go' + Inf construction, inasmuch as the occurrence of the periphrasis is followed by a description of consequences. On the basis of the fact that the utterance continues with a description of the consequence(s) of the action, we can infer that it was actually performed. In Example (41) they did not only think of escaping, but they actually escaped, because, as a consequence, the king's host found the city abandoned. Similarly, in (42) the Turks did not only think of taking out their arrows, but they actually took them out. This is clear from the consequence: an arrow hit and killed a knight called Corberan. Furthermore, as a consequence of the riding described in (43), they caught up with the Cid.

To sum up, the construction *pensar (de)* 'think of' + Inf presents the event in the infinitive as a completed, thoroughly performed action, which has considerable consequences and, as such, is worthy of the listeners' attention.

2.6.3.4 Començar (a/de) *'begin to'* + Inf

The *començar (a/de)* 'begin to' + Inf is a typical, widely used inchoative periphrasis in both medieval Catalan and Spanish. Its use seems to fulfil a strategy of attracting the listener's attention, similarly to other inchoative constructions described in Detges (2004) (cf. Section 2.3.3). As Table 15 shows, the finite verb and the infinitival complement is linked together in most of the cases with the preposition *a* in Catalan, although the preposition *de* is also used relatively frequently. Moreover, there are some sporadic cases of the phrase without a preposition. I regard all these morphological realizations as stylistic variants, because their documented uses do not reveal any difference in meaning. The situation in Spanish is very similar, but here the proportion of the prepositions *a* and *de* are almost the same, the variant without a preposition being the rarest (see Table 16). Different variants can appear within the same document. In the case of the coordination of two or more infinitives, they are sometimes introduced by the same preposition, but frequently the infinitives are introduced by different prepositions. There are several possibilities: it may occur that the preposition *de* introduces the first infinitive, while the preposition *a* introduces the second, or vice versa, but sometimes the first infinitive is introduced by a preposition, but the second appears without it.[52] The distribution of occurrences of *començar (a/ de)* 'begin to' + Inf is shown in Tables 15 and 16 in the Catalan and Spanish texts of the corpus, respectively.

Table 15. Distribution of occurrences of *començar (a/de)* 'begin to' + Inf in Catalan texts of the corpus

		Jau	Desc	Munt	Per	Ger	
començar a + Inf	Preterit	6	34	31	4	45	
	Present/preterit syncretism	-	-	2	-	-	122
començar de + Inf	Preterit	23	13	2	1	5	
	Present/preterit syncretism	7	-	-	-	-	51
començar + Inf	Preterit	-	-	2	-	1	3
començar (mixed) + Inf	Preterit	-	-	4	-	8	12
Total		36	47	41	5	59	188

Table 16. Distribution of occurrences of *començar (a/de)* 'begin to' + Inf in Spanish texts of the corpus

	Cid	Mar	Tres	GrCo	Alf	Tris	Arn	Pier	CrPC	
començar a + Inf	1	3	-	72	7	128	9	29	36	285
començar de + Inf	8	3	2	56	1	112	-	3	86	271
començar + Inf	-	1	-	7	1	1	1	2	4	17
començar (mixed) + Inf	-	-	-	38	-	40	3	10	16	107
Total	9	7	2	173	9	281	13	44	142	680

In the following, let us consider the most frequently used infinitives in the BEGIN TO + Inf construction. The verb *començar* is often combined with infinitives of communication (e.g. the Catalan *parlar* 'to speak, to talk', *dir* 'to say, to tell', and *cridar* 'to shout, to cry' or the Spanish *fablar* 'to speak, to talk', *dezir* 'to say, to tell', *maldezir* 'to swear, to curse', *demandar* 'to ask', *preguntar* 'to ask', *contar* 'to narrate, to tell (a story)') and with those of mental/emotional state/act (e.g. the Catalan *plorar* 'to cry', *riure* 'to laugh', *fer dol* 'to lament, to moan', or the Spanish *llorar/hazer llanto* 'to cry, to lament', *pensar* 'to think', *cuydar* 'to think', *rreyr* 'to laugh', *sonreyrse* 'to smile'). In addition, dynamic verbs also enter the periphrasis, among them infinitives describing sudden, single actions (e.g. the Catalan *ferir* 'to attack, to hit, to wound' and *tirar* 'to throw', or the Spanish *ferir(se)* 'to attack, to hit, to wound (each other)' and *tirar* 'to throw'). Verbs of motion (e.g. the Catalan *fugir* 'to flee, to get away', *pujar* 'to climb, to come up', and *exir* 'to go out', or the Spanish *foyr* 'to escape' and *ir* 'to go', or *andar* 'to walk, to go') form a subgroup of dynamic verbs. Naturally, verbs of durative events or states (e.g. the Catalan *cavar* 'to dig, to trench' or the Spanish *rreynar* 'to reign' and *guerrear/fazer guerra* 'to war') can also enter the periphrasis formed with *començar*, where the construction indicates the beginning of that event or state.

About a quarter of the occurrences in Catalan and more than half of them in Spanish belong to two semantic groups: infinitives of communication and those of mental/emotional state/act. Therefore, these are the collocations which should be examined with more attention. A typical use of *començar (a/de)* 'begin to' + Inf is where the finite verb 'begin' is combined with the infinitive of communication *dezir* 'to say, to tell'. In the following section I present the use of this frequent collocation.

a. Començar (a/de) dezir *'begin to say / to tell'*

The phrase *començar (a/de) dezir* is the most frequent combination, which appears in the Catalan texts of the corpus 8 times and in the Spanish texts 57 times.

In the first group of examples the occurrence of *començar (a/de) dezir* is followed by a more detailed description of the content of the communication in a direct quotation form, as in the Spanish example in (44):

(44)

E ordoño començo   de-le   dezir.   Don diego gonçales
       begin.3Sg.PerfP Prep-Clit tell.Inf

tomad este cauallo.

'and Ordoño **began to tell him**. Don Diego Gonçales, take this horse.'

(CrPC 87v)

This group contains most of the occurrences and suggests that the function of the periphrasis consists in addressing attention to the content of the utterance which follows the periphrasis.

In another group of occurrences, the content of the communication act is not described in detail, but is expressed by the direct object of the infinitive, as in the following Spanish example (45):

(45)

començo   a   dezir   los bienes & noblezas y virtudes que eran
begin.3Sg.PerfP Prep say.Inf

eneste jouen cauallero.

'he **began to list** the good qualities and nobleness and virtues which this young knight had.'

(Pier 22v)

Finally, the third group of occurrences expresses the content of communication in indirect speech, as in the Spanish example (46):

(46)

Et començaron   a   dezir   en su lenguaie. Que aquella yent. non era
   begin.3Pl.PerfP Prep tell.Inf

si non de fierro.

'and they **began to tell** in their language that those people were surely made of iron.'

(GrCo 25r)

The distribution of these uses is shown in Table 17.

Table 17. Distribution of contexts in the use of *començar (a/de) dir / dezir* 'begin to say/to tell'

	Catalan *començar (a/de) dir*	**Spanish** *començar (a/de) dezir*	
Quotation	5	35	40
No content	3	13	16
Indirect speech	-	9	9
Total	8	57	65

Similar occurrences appear with other infinitives of communication as well. Let us consider the following Spanish (47) and Catalan (48) examples.

(47)

E el rrey le començo   a   demandar cauallero donde soys o de qual
      begin.3Sg.PerfP Prep ask.Inf

parte venjades.

'The king **began to ask** him: knight, where are you from or where did you come from?'

(Tris 6v)

(48)

E tuit començaren   a   cridar: – Ben diu l'almirall!
   begin.3Pl.PerfP Prep cry.Inf

'And all **began to cry out**: "The admiral says well."'

(Munt I 127,34–36)

Examples (47) and (48) show typical uses of *començar (a/de)* 'begin to' + Inf, where the finite verb 'begin' is combined with a verb of communication and the occurrence of the periphrasis is followed by a more detailed description of the content of communication in a direct quotation form. In Examples (47) and (48), the lexemes *demandar* 'to ask' and *cridar* 'to cry' could also represent a single, perfective event. However, in the inchoative periphrasis they acquire a repetitive character. Examples (47) and (48) can therefore be interpreted as describing repeated questions and cries, respectively (cf. also Example (53) in Section 2.6.4.2).

It seems to be reasonable to suppose that the function of the periphrasis consists in addressing attention to the content of the utterance which follows the periphrasis. That is, the inchoative aspect serves as a tool for organizing the discourse, by indicating the beginning of an act of communication. However, it is not the beginning which is important but the continuation, that is, the content of the communicative act.

Another semantic group of infinitives co-occurring with *començar* 'begin' are motion verbs. Verbs of motion, a subgroup of dynamic verbs, also enter the GO- and THINK OF-periphrases. However, motion events can be represented in different ways. Consider the Catalan example in (49).

(49)

E el rey, qui assò hac entès, va pendre ses armes, e muntà a caval e

comensà   *a*   *córrer*   *aprés d'éls; e no foren pus de ·XX· cavalers ab él,*

begin.3Sg.PerfP   Prep   run.Inf

que·ls altres no eren tantost aparelats. El rey era molt bon cavaler e coratgós e era molt bé encavalquat, sí que·ls altres cavalers no podien tant córrer com él; sí que·l rey fo molt a davant de tota sa companya e conseguia en aquels qui se'n fugien, sí que·ls fo molt prop, e regiraren-se e conegren que aquest era·l rey.

'And the king, who got to know that, took his arms, and mounted his horse and **began to ride** at full speed after them; there were no more than 20 knights with him, because the others could not get ready soon enough. The king was a very good knight and courageous and could ride very well. So the other knights could not ride so fast as him; so that he rode quite ahead of all his companions and was about to catch up with those who were escaping. So he was close to them and they turned round and recognized that he was the king.'

(Desc II 43,5–15)

The Catalan Example (49) shows an occurrence of *començar (a/de)* 'begin to' + Inf with an infinitive of motion. The event described in the infinitive is extended in time: the act of the king's riding is presented as prolonged in time. However, the broad context[53] shows the king's riding on his horse as 'frozen', because the storyteller stops and describes how it was performed, what the circumstances were, what events occurred during this action etc. The event wedged in is a very considerable one: the enemy recognizes the king's person. This recognition is considered to be of great significance by the storyteller, who wants to emphasize it by means of 'preparing' it through the use of the inchoative periphrasis.

To sum up, the investigation of occurrences of *començar (a/de)* 'begin to' + Inf in their broad contexts sheds light on the fact that this construction serves as a tool for emphasizing subsequent events. Events described in the inchoative periphrasis serve as a frame, which makes it possible to wedge in events considered to be more important. Events described in lexically perfective infinitives will be reinterpreted in the construction as extended in time, and in this way they can be interrupted by another event. With this kind of technique the storyteller manages to highlight embedded events and to relegate the event described in the infinitive of the BEGIN TO-construction to the background. By freezing the picture of the event at its beginning, the storyteller directs the audience's attention to the subsequent part of the story.

2.6.4 A comparison of the constructions with 'go', 'think', and 'begin'

2.6.4.1 Infinitive types in the periphrases with the verbs 'go', 'think', and 'begin'

Dynamic infinitives enter all three constructions, and these combinations mostly appear in descriptions of battle scenes (e.g. the Catalan *ferir* 'to attack, to hit, to wound' or the Spanish *ferir* 'to attack, to hit, to wound', *conbatir* 'to combat / to assault (a city)'). Verbs of motion (e.g. the Catalan *fugir* 'to flee, to get away' and *exir* 'to go out', or the Spanish *foyr* 'to escape' and *ir* 'to go') form a subgroup of dynamic verbs. All three periphrases can be used with the same motion – and other dynamic – infinitives, i.e. these are the occurrences which show the highest degree of similarity. It is precisely on the basis of uses of BEGIN TO + Inf and GO + Inf with dynamic infinitives that Detges (2004) supposes that GO + Inf is a representative of the inchoative aspect. Therefore, these are the contexts which should be examined with more attention. In the following I present a comparison of the constructions by the qualitative analysis of these contexts. Let us begin with collocations with the infinitive *ferir* 'to attack, to hit, to wound', which represent a deeply entrenched schema in other medieval Romance languages as well.

2.6.4.2 Anar ferir/ir (a) ferir *'go to attack/to hit'*, pensar (de) ferir *'think of attacking/hitting'*, començar (a/de) ferir *'begin to attack/to hit'*

The infinitive *ferir* 'to attack, to hit, to wound' is one of those infinitives that appear in all three periphrases, forming a frequently used typical phrase in the medieval language state, not only in Catalan, but also in Spanish and in French. Because of this, it is worth comparing occurrences of this infinitive in all three constructions. The combination of the infinitive *ferir* with the verb GO corresponds to an entrenched schema, while with THINK and BEGIN there are fewer occurrences. In sub-section *a* of Section 2.6.3.2, I have already presented the contexts of use of the constructions *anar ferir* and *ir (a) ferir* 'go to attack, to hit, to wound'. Now I examine occurrences of *ferir* in the THINK OF-construction. The distribution in the Catalan corpus is shown in Table 18.

Table 18. Distribution of contexts of the construction *pensar (de) ferir*

	Munt
Consequence	7
Description	1
Total	8

The collocation *pensar (de) ferir* only appears in Muntaner's Chronicle, and it is used in similar contexts to those of *anar/ir (a) ferir* presented above: the subsequent utterances describe consequences of the action described in the infinitive. In this way, an inference of the actual performance of the action emerges, as in (50).

(50)

*E així con havíem fet en les altres batalles, **pensam tuit de ferir** ensems en ells, de cavall e de peu; així que plac a nostre senyor ver Déus que los vencem.*

'And as we had done in the other battles, we **attacked** (= **thought of attacking**) them all together, horse and foot. So it pleased Our Lord the true God that we should vanquish them.'

(Munt II 96,16–19)

As for the construction *començar (a/de) ferir*, there are only a few occurrences in the Catalan texts, while in Spanish it is more frequent. The distribution in the corpus is shown in Table 19.

Table 19. Distribution of contexts of the construction *començar (a/de) ferir* 'begin to attack'

	Catalan *començar (a/de) ferir*		**Spanish** *començar (a/de) ferir*			
	Desc	Munt	GrCo	Tris	CrPC	
(Description +) consequence	2	-	2	5	-	9
Embedded event, difficulty	1	1	1	17	1	21
Total	3	1	3	22	1	30

The contextual analysis reveals that in most of the cases either the objects of the attack or the subjects are in the plural: *ferir en ellos* 'to attack them', *ferir en los moros* 'to attack the Moors', *(todos) ferir en el* '(all together) attack him'. As a consequence, the event of attacking seems to be represented as divided into several sub-acts, instead of a momentary, punctual event. Such a representation is often confirmed by certain adverbs, such as *(ferir) a diestro & a sinjesto* '(to attack) right and left' or *(ferir) a vna parte & a otra* '(to attack) on one side and on the other'.

There are only a few exceptions, and occurrences where there is an object in the singular and perfectivity can be inferred – as occurs in the case of the GO-periphrasis (cf. Example (28) in Section 2.6.3.2) – are completely missing. In the case of the few examples with an object in the singular, the periphrasis highlights the beginning of the event and the context does not allow for an inference of perfectivity, as can be observed in the following Spanish example:

(51)

et luego **començo a ferir** *el cauallero contra tristan que esta ya aparellado de iustar et desque tristan lo uio uenir …*

'then the knight **began to attack** Tristan, who is ready to fight, and when Tristan saw him
   coming …'

(Tris 111v)

The event of attacking described in the infinitive in (51) is presented as one which was begun but was not yet completely performed: the subsequent clauses make it clear that the knight who intended to attack Tristan really started to move, which could be understood as the beginning of the attack, but had not yet reached him. The periphrasis in (51) describes the beginning of a single act of attacking. More typical uses are where the context allows for a conceptualization of the act described in the infinitive as several acts of attacking performed in succession or simultaneously. In these cases, the consequences of the attack are sometimes described in the context as well, thus the perfectivity of the attack can be inferred.

(52)

& començo de ferir en los diez caualleros a diestro & a sinjesto & fizo tanto & en tal manera que en poco rrato non fjnco cauallero njnguno enla plaça delante el que todos non fueron muertos o feridos

'and he **began to attack** the ten knights left and right, and he did it so much and in such
   a manner that in a short time there was no knight who could stand up before him in the
   square, they all were killed or wounded'

(Tris 37v)

In contrast to the typical contexts of the use of 'go' + Inf and 'think' + Inf, in the use of the construction 'begin to' + Inf a new context type appears: the act of attacking begins, but it is immediately interrupted by another event. In this way the storyteller relegates the event described in the infinitive of the BEGIN TO-construction to the background and the attention is directed towards a new event. This context type is very frequent, containing more than two-thirds of the occurrences. Let us consider an example.

(53)

*[el rey] fuese contra los escuderos & **començo**   los **de**   **ferir***

                                start.3Sg.PerfP   Prep attack.Inf

*& la donzella que auja traydo a tristan vio commo la falsa donzella se acogia & **començo** **a dar bozes** contra tristan & **a dezjr** Señor catad que non se vaya la mala muger ca por ventura otro tal faria ella otra vegada tristan ferio el cauallo contra el castillo & ante que ella entrase tomola por los cabellos & pusola en poder del Rey & tornose contra los escuderos*

'[the king] went to the squires and **began to attack** them. And the girl who had taken Tristan there saw that the false girl was about to go away and **began to shout** to Tristan and **to say**: Sir, look out! Don't let this bad woman escape, because maybe she would act in the same way again. Tristan spurred his horse and went towards the castle and before she was able to enter, he took her by the hair and handed her over to the king. Then he turned towards the squires.'

(Tris 131v)

The phenomenon presented above can be observed in the Spanish Example (53): due to the object in the plural (*los* 'them', i.e. the squires), the act of attacking described in the infinitive can be understood as a series of multiple acts of attacking. The periphrasis expresses the beginning of this series of attacks. As mentioned above (in Section 2.3.3) inchoative constructions are compatible with durative verbs, which denote actions or processes characterizable by a time interval, and whose left starting point is signalled (Kiefer 2006: 169–170). The act of attacking in itself is a punctual, perfective event. However, the representation of the event of attacking changes when it occurs in the inchoative periphrasis: instead of a momentary, punctual event, it is represented as a durative one. The broad context of (53) highlights that it is interrupted by another, unexpected event. The attention is directed from the attack of the squires towards a new event: a dishonest person, who was the cause of various troubles before, is about to escape. Keeping her back from escaping is more urgent than disarming the squires, which is delayed. The representation of the event described in the infinitive has been begun but it has been interrupted, and the inchoative aspect in this example seems to fulfil the function of directing the attention to that part of the story which follows the BEGIN TO + Inf periphrasis. That is, it is not the event described in the infinitive which the speaker seems to highlight, but a subsequent event embedded in the narrative.

2.6.4.3 Anar dir/ir (a) dezir *'go to say/to tell'*, començar (a/de) dir/començar (a/de) dezir *'begin to say/to tell'*

The infinitive *dir/dezir* 'to say, to tell' enters both the BEGIN TO + Inf and the GO-construction. I have already presented the use of the collocation *començar*

Table 20. Distribution of contexts of the constructions *anar dir* and *ir (a) dezir* 'go to say/to tell'

	Catalan *anar dir*			Spanish *ir (a) dezir*		
	Desc	Munt	Per	Ger	CrPC	
Content unexpressed	2	1	2	-	2	7
Indirect speech	-	-	1	3Prep	-	4
Direct quotation	-	-	3	-	-	3
Total	2	1	6	3	2	14

Table 21. Distribution of contexts of the constructions *començar (a/de) dir* and *començar (a/de) dezir* 'begin to say/to tell'

	Catalan *començar (a/de) dir*			Spanish *començar (a/de) dezir*					
	Jau	Munt	Ger	GrCo	Tris	Arn	Pier	CrPC	
Content unexpressed	2	1	-	-	7	-	1	5	16
Indirect speech	-	-	-	2	1	1	-	5	9
Direct quotation	2	-	3	-	18	6	3	8	40
Total	4	1	3	2	26	7	4	18	65

(a/de) dir/dezir. We have seen that in typical uses of *començar (a/de)* 'begin to' + Inf the finite verb 'begin' is combined with a verb of communication and the context contains a detailed description of the content of communication in a direct quotation form. In these cases, it seems to be reasonable to suppose that the function of the periphrasis consists in addressing the attention to the content of the utterance which follows the periphrasis. Although there are some occurrences with the verb *anar/ir*, the use of a communication infinitive in the GO-construction does not seem to be typical. Occurrences where the phrase *anar dir* is followed by the quotation of the content of the communication act are in the present tense and seem to be grammaticalized forms (see Example (38) above).[54] However, this perfective past seems to be pragmatically marked in contrast to the simple past, because it can appear in a limited range of contexts at that time. For occurrences with the finite verb in the preterit, the motion-then-action reading is always accessible.

2.6.4.4 Infinitives describing dynamic, punctual actions

In this section I compare the uses of our three constructions with infinitives describing dynamic, punctual actions. Although these uses cover a wide range of infinitives, these are the contexts which should be examined with more attention, because this type of collocation is typical in the case of GO + Inf, and already Colon (1959/1978: 130) notes that before 1350 only these lexically perfective infinitives could enter the construction.[55] Example (54) shows a Spanish occurrence.

(54)

*fueron todos **tomar** sus armas por entrar dentro en la uilla. Mas el maestre del Temple que dezien don bernalt de Tremolay salio adelante por defender que ninguno non entrasse si non sus freyres.*

'all of them **went to take** their armour in order to get inside the villa. But the Grand Master of the Templars, called Bernalt de Tremolay came out to prevent anyone from entering, except his brothers.'

(GrCo 68v)

In Example (54) the infinitive *tomar* 'to take' describes a punctual action. In the context of use, an inference emerges according to which this action was thoroughly completed, since – probably with their armour – they went to the villa. A similar perfectivity can also be observed in the uses of *pensar* 'think' with infinitives of dynamic, punctual actions. In the following discourse (presented in Section 2.6.3.3 as (42) and repeated here for convenience as (55)) the Turks not only thought of taking out their arrows, but actually took them out. This is clear from the content of subsequent clauses: an arrow killed the knight called Corberan.

(55)

*e los turcs, qui veeren que els muntaven darrera, **pensaren**   ab les sagetes*

think.3Pl.PerfP

de   trer,   e per desastre una sageta va ferir En Corberan, qui s'hac

Prep   take.out.Inf

desarmat lo cap per calor, per lo pols. E aquí ell morí.

'But the Turks, seeing them come up after them, **took out (= thought of taking out)** their arrows. And unfortunately an arrow hit En Corberan, who had taken off his iron cap because of the heat and dust, and there he died.'

(Munt II 77,2–5)

In contrast to Examples (54) and (55), the fragment in (56) reveals a distinct behaviour of the verb *començar* combined with infinitives describing dynamic punctual actions:

(56)

*& el commo los vio **començo a poner** su escudo delante & **aparejarse** para la justa mas commo fue çerça dellos luego conosçio al rrey & descendio del cauallo & fue le a besar la mano & dixo …*

'and when he saw them, he **began to put** his shield in front of himself and **to prepare** himself for battle. But when he got near them, he recognized the king and he got off his horse and went to kiss his hand and said …'

(Tris 78v)

Although in this fragment the infinitive also describes a punctual, perfective event, we have to change the representation in this context. In the Spanish Example (56), the act of taking the shield and preparing for battle are interrupted and are not completed, because the knight realizes that the person he is going to attack is the king himself, and he changes his mind. He renounces combat and, instead, decides to kiss his hand. By the use of *començar (a/de)* 'begin to' + Inf, the wedged-in event, i.e. the recognition of the king, seems to be emphasized.

A similar change in the conceptualization of the act described in the infinitive can be observed in the Spanish Example (58), also containing the inchoative 'begin to' + Inf construction. A comparison with Example (57), which also originates from a Spanish text, reveals differences in contexts.

(57)

*& **fue dar** a esmerol tan grand **golpe** que le tajo el yelmo*

'and he **gave** (= **went to give**) Esmerol such **a blow** that he cut his helmet'

(Tris 5v)

(58)

*& tristan la oyo mas non dixo palabra & **començo de dar** los **golpes** muy fuertes a galeote aqui veredes dar golpes & rresçebir en tal manera que los caualleros que guardauan la batalla dezian que de grand fuerça son los dos caualleros*

'Tristan heard her but didn't say anything and **began to give** Galeote very hard **blows**. And so you can see giving and receiving such blows that all knights looking at the fight said how strong both knights are'

(Tris 21r)

While in (57) a momentary, punctual act is presented, followed by the description of consequences, in (58) the act described by the infinitive acquires a repetitive character in the inchoative periphrasis. In the subsequent clauses, the storyteller elaborates a picture of giving blows as an act repeated several times.

Since the periphrasis *començar (a/de)* 'begin to' + Inf refers to the beginning of actions, infinitives of prolonged events can also appear in it.

(59)

*E puis **comensaren a cavar** tro al mur major, e els sarraÿns, qui ho conegren, feeren altra cava dedins endret d'éls; sí que s'encontraren ensems e aquí agren gran batayla, sí que·ls crestians agren la cava a desemparar.*

'and then they **began to dig a tunnel** towards the major wall, and the Saracens, who noticed them, did another tunnel from inside towards them; so they met and they had a great battle, so that the Christians had to leave the tunnel.'

(Desc II 126,26 – 127,3)

In the case of Example (59), as in other similar examples, the act described in the infinitive is interrupted by another, unexpected event.

In contrast, in the *anar ir (a)* + Inf construction, even prolonged events are represented as completed, as in the Catalan example in (60).

(60)

*E, quan vench al matí, hoïdes les misses, **anaren-se armar** los escuders e gran partida dels cavalers; e meteren-se de la part del sas e començaren d'entrar en la vila a peu.*

'And when the morning came, after mass the esquires and many of the knights **went to put on** their armour and worked in on the side next to the camp and began to go into the town on foot.'

(Jau 201)

Although putting on armour is an event extended in time, in the context of Example (60) the subsequent description makes it clear that it had been completed.

2.6.5 A comparison of *començar (a/de)* 'begin to' + Inf, *pensar (de)* 'think of' + Inf, and *ir (a) / anar* 'go' + Inf

The analysis of historical occurrences has led us to the finding that the constructions *començar (a/de)* 'begin to' + Inf, *pensar (de)* 'think of' + Inf and *ir (a) / anar* 'go' + Inf are functionally similar to each other: all three are appropriate to realize a strategy which aims to attract the listener's attention. However, as the qualitative analysis of contexts has suggested, the meanings of these functionally similar constructions are different.

The construction GO + Inf characterizes the action or event as a whole, without extension in time, and emphasizes the action itself, mentioning its consequences. This perfectivity in contexts of use might have played a decisive role in the semantic development of the Catalan *anar* + Inf construction: it might have been integrated into its coded meaning.

The THINK OF-periphrasis has an intentional meaning and emphasizes the dynamism of the action by referring to intentionality, i.e. the subject's volitional involvement in the action. Historical data have reflected a similarity of its contexts of use to those of the GO-construction, since they both emphasize the action described in the infinitive. The subsequent utterances in the context often describe the consequences of the action described in the infinitive, which suggests that this construction also presents the event in question in its entirety, as a completed action, which takes place as a whole, without being interrupted.

BEGIN TO + Inf is a typical inchoative construction, which seems to suggest cognitive relevance, as Detges (2004) describes it. Although its contexts of use at first sight may seem to be similar to those of the GO- and BEGIN TO-constructions, a deeper contextual analysis has revealed that the construction formed with *començar* 'begin' has only a 'preparatory role'. It describes the action or event as extended in time and, through the reference to its beginning, draws the listener's attention to the event/s that interrupt/s it.

Differences in uses of the periphrases *anar* + Inf and *començar (a/de)* 'begin to' + Inf revealed by the consideration of broad contexts suggest that, in contrast to what Detges (2004) claims, the occurrences of *anar* 'go' + Inf are not representatives of the inchoative aspect. At the same time, the similarity between the uses of *ir (a)/anar* 'go' + Inf and *pensar (de)* 'think of' + Inf confirms the hypothesis that at a certain stage of its grammaticalization, GO + Inf had a conceptual affinity with THINK OF + Inf, which can be attributed to the salience of the meaning element 'intention' in the semantic structure of GO. At this stage the construction expressed the intention of the subject and not an 'abrupt start' (i.e. inchoative aspect; the adjective *abrupt* is added by Detges 2004). The THINK OF- and GO-constructions both emphasize the action in the infinitive, by referring to the subject's volitional involvement, and present it as the culmination of the story. Figure 1 visualizes how the speaker tries to capture the addressee's attention through the use of BEGIN TO + Inf, THINK OF + Inf or GO + Inf.

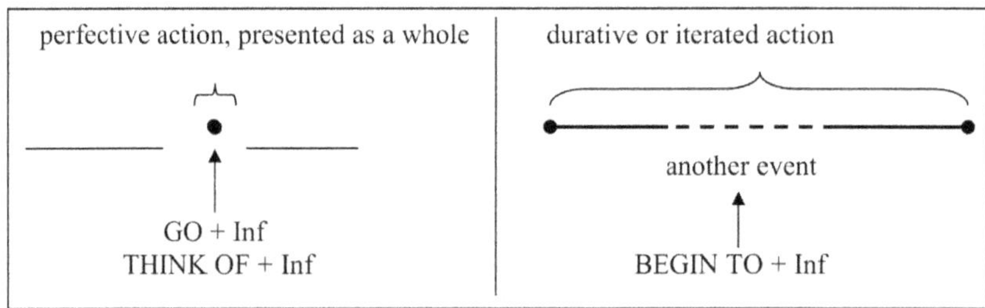

Figure 1. Strategies of capturing the listeners' attention with the use of GO + Inf, THINK OF + Inf and BEGIN TO + Inf

The verbs *començar* 'begin', *pensar* 'think', and *ir/anar* 'go' all occur with similar types of infinitives, but they add different nuances to the description of the event. In the cases of the GO- and THINK OF-periphrases, the act described in the infinitive is highlighted and is referred to as a whole, a completed event with considerable consequences. In the case of the BEGIN TO-construction the event in the infinitive is described as a durative one, but one which is interrupted as soon as it has begun. With this strategy the subsequent wedged-in events are emphasized.

To sum up, the comparison of these three medieval periphrases in Catalan and Spanish has revealed a similarity between uses of *ir (a)/anar* 'go' + Inf and *pensar (de)* 'think of' + Inf, and a difference of both from the uses of *començar (a/ de)* 'begin to' + Inf. On the basis of these observations, we can conclude that, at a certain stage of its grammaticalization, GO + Inf had a conceptual affinity with THINK OF + Inf, which can be attributed to the salience of the meaning element 'intention' in the semantic structure of GO. At this stage the construction expressed the intention of the subject and not the inchoative aspect. Since 'intention' is a content that originally forms part of the meaning of the lexical item GO, we can assume that it may not have emerged as an implicature but has become salient within the meaning of GO in frequent contexts of use. Through the use of the periphrasis, the speaker may have intended to refer to the agent's involvement in an event considered to be the turning point of the story.

Additional evidence comes from the finding that in medieval Spanish and Catalan the *ir (a)/anar* 'go' + Inf construction often contains infinitives which refer to emotional events, like *abraçar* 'to embrace' or *besar* 'to kiss'. The hypothesis of an 'abrupt start' does not fit with these occurrences. They are not characterized by abruptness, neither do they direct the attention to the subsequent utterances. Rather, they are emotionally charged scenes, which suggest the subjects' active involvement. Compare the following examples from medieval Catalan (61) and Spanish (62).

(61)

Madona la reina e sa germana **anaren-se**   **abraçar;**   *e estegren així*
                              go.3Pl.PerfP-Clit embrace.Inf

abraçades, e besant e plorant, que null hom no les podia partir.

'My Lady the Queen and her sister **went to embrace/embraced** each other and remained thus embracing and kissing each other and weeping much, so that no one could part them.'

(Munt I 181,32–34)

(62)

El Çid a doña Ximena **íva-la**   **abraçar,**
                    go.3Sg.ImpP-Clit   embrace.Inf

'Cid **was embracing** (= **was going to embrace**) doña Ximena

doña Ximena al Çid la mánol' va   **besar,**
                    go.3Sg.Pres kiss.Inf

doña Ximena **kissed** (= **goes/is going to kiss**) his hand

llorando de los ojos, que non sabe qué se far
weeping from her eyes, she does not know what to do'

(Cid 368–370)

Examples (61) and (62) represent emotionally charged scenes, as is also confirmed by their broad contexts. We can suppose that such a nuance in meaning regarding the subject's emotional involvement can be derived from 'intention', because this semantic component emphasizes a relationship with the 'internal', emotional, and intentional world of the person who carries out the action. Instead of conveying the motion meaning with the use of *ir* + Inf, the speaker seems to emphasize that the action described in the infinitive was performed in a non-ordinary way: it was more dynamic, more voluntary and more emotional than our actions usually are.

2.6.6 A further argument in support of the preterit hypothesis

In cases in which different manuscripts or translations of the same text have been preserved, the broadening of our data sources is possible. The comparison of manuscripts originating from different language stages may provide useful information about language change, because sometimes it is precisely the changing elements of a language which are affected by variations in different manuscripts. Since copyists probably intend to write a text which is also comprehensible for contemporary addressees, it may occur that when copying texts foreign to them, they change some linguistic forms which are considered to be archaic. Where variations occur, we can interpret these differences as signs of the copyist's linguistic intuition and, as such, we should proceed cautiously when interpreting them.

Contexts of use of the Catalan construction *anar* + Inf are highly constrained in the earliest historical documents, thus it is especially welcome to also have a later change in a textual location relevant for our study. Bruguera (1981: 38–40) compares various manuscripts of the Chronicle of James I (namely, H Poblet 1343, C 1380, D 14th century, E 15th century, 1557). The earliest manuscript originates from 1343, and the latest from 1557: this two-century difference allows us to acquire useful information about the history of the construction affected by the change. As we have already seen (Section 2.5.2), at corresponding places in different manuscripts the same construction may appear with or without a preposition. Here I will discuss another interesting difference, this time concerning the tense of the finite verb. As I have mentioned in Section 2.3.7, some authors start from the present tense variant, while others from the preterit tense variant when analysing the semantic change of the GO-construction. This interesting change affecting the tense of the finite verb in different manuscripts provides additional evidence to decide between these rival hypotheses.

The change is described in Bruguera (1981: 39–40) in the following way. The earliest manuscript (called H Poblet, from 1343) contains the following variant:

(63a)

*E cridàvem: 'Vergonya, cavalers!' E **anam-los***   *ferir*

                         go.1Pl.Pres/PerfP-Clit   attack.Inf

e esveÿm-los.

'And we were shouting: "Shame, knights!" And **we go/went to attack/attacked** them and broke their line.'

In later manuscripts a slightly different form appears at the relevant place:

(63b)

*E cridàven: 'Vergonya, cavalers!' E **van-los***   *ferir*   *e esvaÿren-los.*

                         go.3Pl.Pres-Clit   attack.Inf

'And they were shouting: "Shame, knights!" And **they go to attack / attacked** them and broke their line.'

The change consists in altering the grammatical number and person of the verbal forms: the later copyist uses the 3Pl form instead of the 1PL form of the earliest manuscript. This is relevant for us, because the 1Pl form of *anar* is characterized by a present–preterit syncretism, which is not true of the 3Pl form. The 3Pl form *van* in (63b) is conjugated unambiguously in the present, while in (63a) it cannot be revealed with certainty whether the form *anam* was intended as a present or as a preterit form. Accordingly, the collocation *anam(-los) ferir* in (63a) can be interpreted in two different ways:

(I)   If the verb *anar* is intended to convey its literal motion meaning:

      (a)   it can be a historical present verbal form, with the meaning 'we go to attack them', or

      (b)   it can be a preterit verbal form, with the meaning 'we went to attack them'.

(II)   If the verb *anar* is not intended to convey its literal meaning, but functions as an auxiliary, that is, we are dealing with an analytic past tense (or an aspectual periphrasis):

      (a)   it may be the case that the speaker intended the form *anam* as a morphologically past form, with the meaning 'we attacked them' (or with an aspectual meaning), or

      (b)   it may be the case that the speaker intended the form *anam* as a morphologically present form, with the meaning 'we attacked them' (or with an aspectual meaning).

In cases (IIa) and (IIb) the meaning is the same; there is only a formal difference. The cases (Ib) and (IIa) must be excluded if we accept the historical present hypothesis. However, if we accept the hypothesis according to which the semantic change affected the preterit variant, and the present tense version is a later development due to analogical changes (see Juge's (2006) hypothesis in Section 2.3.6), we can accept the cases (Ib), (IIa), and (IIb), but they would belong to successive stages of grammaticalization, (Ib) preceding (IIa), and (IIa) preceding (IIb).

Let us begin with Bruguera's (1981) explanation of the occurrences. Bruguera claims that occurrences of *anar* + Inf in the Chronicle of James I do not have a past tense value, so they are not representatives of the analytic past tense. Bruguera (1981: 31) draws attention to the observation that beyond these morphologically ambiguous forms there are no historical present forms in the text. Relying on this observation, he regards these occurrences as preterit forms. In contrast, he interprets the two occurrences of *anar* + Inf where the finite verb is conjugated unambiguously in the present as historical present forms. As for the form *anam*, he accepts the hypothesis in (Ib), that is, he interprets the form in question as the preterit, conveying the source meaning, and this reading seems to be acceptable to me, too. Since, except for morphologically ambiguous forms of the *anar* + Inf periphrasis, there are no further historical present forms in the text, it would be counter-intuitive to suppose a historical present use at that location, restricted to this single construction. Relying on a thorough analysis of occurrences, the hypothesis in (II) can be rejected. Although there are occurrences revealing a more advanced stage of grammaticalization,[56] if the literal interpretation is possible, we must attribute it, and this is true also for (63a).

Now let us turn to the later manuscripts. As I have mentioned above, these contain a 3Pl form in the corresponding location, which is revealing, because this grammatical person does not show the present–preterit syncretism: the morphology of the form *van* is unambiguously present. Naturally, we have to take into consideration that the choice of the later copyist does not provide information about what was intended in the original, only about the copyist's own interpretation of the original form. In this sense, the change is a written indication of the copyist's linguistic intuition, which provides information about use of the construction in the later language stage which the copy belongs to. What meaning can we attribute to the utterance in (63b)? Let us discuss possible options:

(a)   If the verb *anar* is intended in its literal meaning, *van* must be a historical present form, with the meaning 'we go to attack'.

(b)   If the verb *anar* is not intended in its literal meaning, *van* as a morphologically present form functions as an auxiliary of an analytic past (or of a kind of aspectual periphrasis), with the meaning 'we attacked'.

Bruguera also raises the question of whether the use of a present tense form in the later manuscripts means that the earliest ambiguous form was interpreted as an analytic past tense or was interpreted as a historical present. He accepts the hypothesis in (a), that is, that the copyist interpreted the form *anam* as a historical present. He argues that, since almost all occurrences of *anar* can be interpreted according to its literal meaning, in that language stage we cannot interpret occurrences of *anar* + Inf as instances of an analytical past tense nor as forms in the present. It seems somewhat contradictory that earlier, when writing about the morphologically ambiguous forms, Bruguera (1981: 31) maintained that the text does not contain any historical present forms, while now (1981: 39–40) he assumes that the collocation *van-los ferir* is in the historical present. Furthermore, his argumentation itself is incorrect, because from the fact that other occurrences of the periphrasis can be interpreted literally, it does not follow that we must interpret every occurrence in this way, since forms corresponding to different degrees of grammaticalization may coexist in one and the same language stage.

As Bruguera explains, the copyist interpreted the form *anam* as the historical present, which is why he changed it to a present tense form. Nevertheless, it is counter-intuitive to assume a historical present usage which affects only two forms in such a long text, and it would be strange if the copyist had not realized that present tense forms are almost completely absent. Unlike Bruguera, I think the hypothesis in (b) seems to be more acceptable. According to this, the copyist interpreted the occurrence in question as an instance of the analytic past tense (even if originally it was probably intended in its literal sense) and, given that in the contemporary language state the present tense forms were already in use, 'translated' the ambiguous form *anam* with a present form.

In summary, the change presented in this section is relevant, because it affects the syncretic form *anam*, interpretable as either conjugated in the past or in the present tense. If we assume that in the original text the form *anam* was intended as a preterit form (which seems to be probable on the basis of other considerations, too) and in that context conveyed the literal motion meaning, then changing it to an unambiguously present tense form in a later copy confirms Juge's hypothesis concerning the evolution of *anar* + Inf. This change is a good illustration of the kind of reanalysis assumed by Juge, namely, that language users reinterpreted *anar* + Inf as a periphrasis with a present tense auxiliary due to the high frequency of the 1Pl form *anam*. This morphologically ambiguous verbal form may have evoked in the addressees both the present and past tense paradigms.

Although the problem of the interpretation of present tense verbal forms has already been addressed, in the following section we will discuss it in more detail in connection with the notion of the historical present.

2.6.7 Historical present

In this section I address two related research questions:

vi.   Are the present tense variants of 'go' + Inf instances of use of the historical present, and if not, what are they?
vii. Should the present or the past tense variant be our starting point when describing the semantic change in the construction?

One of the main differences between the three analyses discussed above (cf. 2.3) consists in the different reasons to which we attribute the shift in the tense of the auxiliary. The first two accounts (i.e. Colon 1959/1978, 1976/1978; Detges 2004) attribute it to a spreading historical present usage, and the third (i.e. Juge 2006) to morphological reasons. However, they all have something in common: they operate with an implicit notion of the historical present; that is, they do not say explicitly what they mean by this term. If we want to argue for or against a historical present usage in medieval contexts of *anar* + Inf, we should have a clear notion of the historical present.

Colon's and Detges's argumentations are based on an intuitive notion of the historical present, and they seem to conceive of it as the present tense used to refer to past events and whose function is to animate the narration. However, the use of the historical present is a more complicated issue. Juge (2006: 319) gives some more information about his notion of the historical present, arguing that present auxiliary versions cannot be representatives of the historical present, because they appear 'with no other present tense forms nearby'. This suggests that historical present forms usually do not appear in isolation. In a later work, Juge (2008) emphasizes the following characteristics of the historical present: (a) verbs formally in the present tense are used with past temporal reference and express foregrounded information, (b) background information is expressed with verbs in the present progressive or the past, and (c) there are extended sequences of verbs conjugated in the present. These properties do not characterize narrative texts containing the earliest occurrences of *anar* + Inf, which argues against a historical present usage in these documents.

I will now attempt to briefly summarize what is worth knowing about the historical present for the purposes of the present discussion. Historical present is traditionally defined as the present tense used in a narration of events set in the past, and it has been suggested that it makes stories more vivid by bringing past actions into the present (see the relevant literature given in Wolfson 1979: 169 and in Schiffrin 1981: 46). Usually it occurs when relating the most crucial events in a narration. Fludernik (1992) points out that the historical present use reflected in early texts is related to an oral pattern of storytelling. Later, however, as the written standard is developed, historical present usage suffers some changes (Fludernik 1992: 1; cf. also Fludernik 1992: 21).

Beyond the simple characterization presented above, some interesting new findings on the historical present have been put forward in the literature from a discourse analytic perspective, which concern the role of the use of the historical present in narrations. The first is Wolfson's (1979) claim that the conversational historical present has no significance in itself. To put it differently, it is not simply the historical present use itself that matters, but the switching between present and past tenses. This present–past alternation is a discourse phenomenon which organizes and segments the narrative (cf. Wolfson 1979: 178): it separates narrative events from each other, and, at the same time, highlights the turning points of a story. Wolfson (1979: 172) also emphasizes that 'many stories are so organized that what seems to be the most important event is given in the past tense'. That is, she does not accept the claim that the historical present has the function of dramatizing certain acts. Rather, as Schiffrin (1981: 59) explains, the shift to the historical present seems to be an 'internal evaluation device: it allows the narrator to present events as if they were occurring at that moment, so that the audience can hear for itself what happened, and can interpret for itself the significance of those events for the experience'. Silva-Corvalán's findings (1983: 774–775) offer Spanish support for Schiffrin's claim for English that historical present usage is an internal evaluation device.

The second finding is that historical present forms tend to occur more frequently in specified points in the narrative but are missing in others. Schiffrin (1981) analyses how the use of the historical present depends on the organization of narratives. Furthermore, Fludernik (1992: 23) emphasizes that the dynamics of the historical present should be discussed in relation to narrative episodes in their totality.

The present auxiliary versions of *anar* + Inf can be found at certain points of the narrative, namely, in the description of events which can be considered turning points of the story. However, we should not conclude from this that we are dealing with a historical present usage. As Colon (1959/1978: 129) points out, both the periphrasis and the historical present share the same function, and this can be the reason why they appear at the same points of the narrative: in the description of events to which the speaker wants to attract the hearers' attention. We cannot conclude that we are dealing with a historical present use because there are at least two other options to explain the presence of the construction at these points of the narration. On the one hand, it can be due to the value of the periphrasis in itself, if we regard it as a pragmatically marked construction at that language state: either as an inchoative or as an intentional periphrasis, it was a convenient way of attracting the listeners' attention to a given part of the text. On the other hand, the present tense of the auxiliary can be also a simple morphological characteristic as described in Juge (2006), i.e. the tense of the auxiliary of an already grammaticalized construction.

It is also worth noting Schiffrin's (1981: 47) finding, according to which in narrative clauses about 30% of the verbs are in the historical present. Moreover, she notes that 'there is a tendency for verbs in the same tense to cluster together … Thus sequences with rapid alternation between the HP [historical present] and the P [past] are not typical' (Schiffrin 1981: 51) (for the clustering of historical present forms, cf. also Pinkster 1990: 73, 75). Investigating the historical present/ preterit alternation in oral Spanish narrative, Silva-Corvalán (1983: 767) finds a very similar percentage for Spanish: '[O]f 476 narrative clauses, 156 (32.7%) verbs are in the historical present'. She claims that 'this cross-linguistic similarity may respond to a universal pragmatic function' (Silva-Corvalán 1983: 767, n. 13). These two characteristics, again, do not fit the use of *anar* + Inf in the present tense in the medieval Catalan chronicles. In these texts the proportion of the present variants of the construction does not reach 30%, and the occurrences do not cluster together. Although in some passages of Muntaner's Chronicle some occurrences appear close to each other (see Soldevila's observation presented in Section 2.3.2), it is not a general characteristic of the use of the periphrasis in all chronicles, and it can also be attributed to a higher degree of grammaticalization of the construction in that language stage. Moreover, what is typical for the use of *anar* + Inf in these medieval texts is precisely the rapid alternation of the present tense versions of the construction with verbal forms in the preterit, which is not typical for a historical present usage. The main finding, however, which argues contra the historical present hypothesis, is that in the relevant contexts only the finite verbs of the *anar* + Inf construction appear in the present tense, and it would be very strange and, moreover, counter-intuitive to suppose such a historical present use, exclusively affecting instances of a single construction.

2.6.8 Analysis of the occurrences in the corpus from the perspective of semantic change

This section addresses the following research question:

> viii. How and in what contexts may semantic change have been initiated?

In order to reveal the development of semantic change, we have to start from the original situation, i.e. from occurrences in which the verb *anar* in the purposive construction conveys its literal meaning of motion. Among these initial contexts there are some characterized by ambiguity: the original meaning is possible, but at the same time, for some reason, another reading is accessible as well. This ambiguity may have later led to the semantic change in the construction. For this reason, I have examined occurrences in the historical corpus from the point of view of whether it is possible to interpret them according to the motion meaning of *anar*, or the construction *anar* + Inf has to be understood in an alternative sense. When guessing the meanings the construction 'go' + Inf may convey, I have relied on the methodological principle provided in Traugott and Dasher (2002/2004: 44), according to which we should attribute to the grammaticalizing linguistic unit the source meaning in all cases in which it results in a meaningful interpretation. From this perspective, I have differentiated between two main groups of occurrences: (a) those where the motion meaning is possible, and (b) those where the interpretation according to the literal 'motion' meaning results in a nonsensical reading. To illustrate these two types of contexts, consider the following occurrences, (64) and (65).

(64)

anà   *pendre*   *comiat del senyor rei, qui era en la ciutat de Lleida.*
go.3Sg.PerfP   take.Inf   leave
'(Conrado) **went to take leave** of the Lord King, who was in the city of Lleida.'

(Munt I 61,18–19)

(65)

un núul se mès sobre nós, e **va'ns**   *cobrir*   *tots d'aigua*
                               Aux.3Sg.Pres-Clit.1Pl   cover.Inf

així con estàvem agenollats.
'a cloud came over us, and **covered us** (= **?goes to cover us**) with water, as we were kneeling'

(Munt II 89,41–42)

In the case of the occurrence in (64), the motion reading is meaningful, since the king is at a different place, thus a change of location was necessary in order to meet him and to take leave of him. Bruguera (1981) notes that in the Chronicle of James I, the verb *anar* figures mostly in its literal meaning and introduces a new scene which actually happens at a different location. According to this, the occurrences in this chronicle all belong to group (a), as is the case in the example in (64). Bruguera claims that if we translate these examples to Modern Catalan, the use of the analytic past does not result in a meaningful interpretation, because this translation does not express a change of scene, breaking in this way the coherence of the narration. In contrast to Bruguera's opinion, I argue that the example in (64) can be interpreted without the explicit expression of the change of scene, since this information can be inferred from the context; nevertheless, the use of the motion verb assigns dynamism to the description.

In contrast, in the case of Example (65) a similar interpretation must be excluded. The presence of the 1Pl clitic is incompatible with the deictic meaning of *anar* 'go', which expresses motion away from the speaker. At the same time, the subject is inanimate, which contradicts the intentionality involved in the meaning of the purposive construction: a cloud obviously cannot move with the aim of doing something.[57] All these considerations confirm that the occurrence in (65) and similar examples belong to group (b). The distribution of these two types of occurrences in the Catalan texts is shown in Table 22.

Table 22. Distribution of supposed meanings in occurrences of the Catalan *anar* + Inf construction

	Jau	*Desc*	*Munt*	*Per*	*Parl*	*EpC*	
Motion possible	39 (~86.66%)	82 (~85.41%)	164 (~70.08%)	1 (~7.14%)	1	-	287
Motion not possible	6 (~13.33%)	14 (~14.58%)	70 (~29.91%)	13 (~92.85%)	-	1	104
Total	45	96	234	14	1	1	391

The distribution of occurrences reveals an increase in contexts over time where the source meaning can be excluded. While in the earliest chronicles we can almost always assign the motion meaning, in Muntaner the original meaning of *anar* already cannot be attributed in almost every third occurrence, and this situation becomes almost exclusive in later texts. These later occurrences can be regarded as grammaticalized forms, and their spread testifies to the ongoing process of grammaticalization. However, the scenario of semantic change can be examined by relying on an analysis of the occurrences belonging to the first group, where the motion meaning is still possible. Let us examine this context type and its variants.

1. Within the group of contexts which favour the motion meaning, further types can be differentiated. It can be observed that in some contexts the motion meaning is the only one, while others reveal ambiguity. Let us begin with unambiguous contexts. Contexts where a location conceived as the goal of the motion is explicitly expressed belong to this type. In these contexts, the verb *anar* unambiguously conveys the meaning 'go', as in (66).

(66)

E puys, quan vench al maytí, **anam**   *hoir*   **la missa**   *a la*

                    go.1Pl.PerfP/Pres   hear.Inf   the mass

esglea major d'Alagó.

'And when morning came, we **went/go to hear mass** in the great church of Alagon.'

              (Jau 22)

In some occurrences, however, the aim of the motion appears only indirectly. When the verb 'go' figures in a purposive construction, we interpret the location indicated as the place of the event described in the infinitive as the end point of the motion, although this location sometimes is not explicitly expressed in the description. Moreover, in some contexts there is no sharp dividing line between the space of the motion and the location of the event described in the infinitive, since the latter is an extension in space of the former. Correspondingly, in Example (67) below, the goal of the motion is the place where the person to be attacked is located, but it is hard to determine where exactly the motion ends and the event of attacking begins. It is, rather, a philosophical question, and it concerns the way the events described in the narration are conceptualized: whether a motion event performed with the aim of attacking is the beginning of the attack or a separate event.[58]

(67)

e aquells de la companya[59] **anaren**   *ferir*   *en ell.*
                                    go.3Pl.PerfP   attack.Inf

'and they of the Company **went to attack** him'

(Munt II 123,30)

The motion event expressed by the GO-periphrasis in (68), inasmuch as we assume that the form *anà* refers to real motion in space, is the continuation of the motion described in the preceding clause by *passar* 'to pass', that is, the use of the verb *anar* is redundant in a certain sense:

(68)

e veem-lo <u>passar</u> denant nós, e **anà**   *s'asseer*   *en sa cadira.*
                                         go.3Sg.PerfP   Refl-sit-down.Inf

'and we saw him <u>pass</u> before us, and he **sat down (= went to sit down)** in his chair.'

(Jau 525)

In contexts similar to that in (68), the discourse remains interpretable even if we eliminate the verb *anar* (cf. 'and we saw him pass before us, and he went to sit down in his chair' vs. 'and we saw him pass before us, and he sat down in his chair'); in this way it allows for a new reading, different from the source meaning. This type of context is assumed to have triggered the semantic change of the construction.

The new interpretation may have been facilitated by other factors, too. Occurrences allowing the motion reading can be grouped in another way. In some contexts, the continuation of the story facilitates an inference to the effect that the event indicated as the aim of the motion was actually performed, although this piece of information is not explicitly expressed, while in other contexts such an inference does not emerge. Observe the following example.

(69)

E no bastà açò, que enans **anaren**   *combatre*   *Nicòtena, e* <u>*preseren*</u>*-la;*
                                                go.3Pl.PerfP   combat.Inf

'And that was not enough; rather they **went to attack** Nicótena and <u>took</u> it,'

(Munt I 102,34–35)

In Example (69), the participants actually moved to a different location, that is, the motion meaning is obvious, but the actual performance of the act described as the aim of this motion is not explicitly expressed. However, we assume the actual performance of attacking the city in order to have a coherent story. We assume the following events when interpreting the discourse in (69): they went to Nicótena to attack it, they attacked it, and, finally, they took the city. In some occurrences the act described in the infinitive is explicitly indicated, but with a different lexeme. Since in these contexts the explicit indication of the perfectivity of the act described in the infinitive is missing, but is included in the interpretation via inference, they promote a strong association between the use of the periphrasis and the inferential content of perfectivity.

2. In another group of occurrences allowing for the motion reading, the movement is minimal, and sometimes the question arises as to whether it is reasonable at all to assume any movement. The frequent use of *anar* + Inf in this type of context can also promote the semantic bleaching of *anar*. Observe Examples (70) and (71).

(70)

E <u>*con foren al peu de l'escala,*</u> *madona la reina e sa germana* **anaren-se**
                                                                        go.3Pl.PerfP-Refl

abraçar;   *e estegren així abraçades, e besant e plorant,*
embrace.Inf

'And <u>when they came down the ladder, at the foot of it,</u> my Lady the Queen and her sister **embraced/(= went to embrace)** each other and remained thus embracing and kissing each other and weeping much'

(Munt I 181,31–33)

(71)

<u>*encontraren-se ab la host dels sarraÿns*</u> *e* **anaren**   *ferir*   *en éls*

                     go.3Pl.PerfP   attack.Inf

<u>'they met the troops of the Saracens</u> and **attacked/(= went to attack)** them'

(Desc II 103,13–15)

In Example (70) the content of the clause which precedes the one containing the verb *anar* provides the information that the participants were at the foot of the ladder, that is, they were presumably very close to each other. However, since in order to embrace each other they needed to get even closer to each other, the motion meaning cannot be completely excluded. In (71), the two groups of soldiers met, that is, they must have been within sight of each other. Nevertheless, in order to attack each other, they must have been even closer, thus the motion reading is possible. Since these occurrences do not include any change of scene, changing the periphrasis for a simple past would result in a coherent discourse. The historical corpus contains several occurrences where acts of motion are performed within a narrow space. In a frequently occurring context type, these movements are performed within an even more limited space than in (70) and (71), and it is reasonable to suppose that in these contexts the aim of the narrator was to achieve a dynamism by the use of the motion verb and not to describe spatial movement. This interpretation appears in previous literature too, not only for Catalan, but for French as well, where the periphrasis is characterized as a certain kind of stylistic tool (cf. Colon 1976/1978: 144). Observe Example (72):

(72)

E lo senyor rei saltà dins, avant, qui era jove e trempat, e *va-li*   *tal* **donar**

                            go.3Sg.Pres-Clit   give.Inf

per mig del cap, de l'espaa, que el capmall que portava no li valc un diner, que entrò

en les dents lo fenè; e puis *va-li*   *trer*   *l'espaa del cap, e*

               go.3Sg.Pres   pull-out.Inf

va   *'n* **ferir**   *altre, que el braç ab tot lo muscle n'avallà en terra.*

go.3Sg.Pres   attack.Inf

'And the Lord King, who was young and spirited, advanced and **gave (= goes to give) him such a blow** with his sword on the middle of his head, that the cap of mail he was wearing was of no use to him, for he was split open to the teeth. Then the King **pulled (= ?goes to pull)** the sword **out** of this man's head, and **attacked (= goes to attack)** another, whose arm, with the whole shoulder, fell to the ground.'

(Munt II 50,3–8)

The fragment in (72) illustrates the typical early context of use of the periphrasis: a battle scene with fast successive movements within a relatively limited space. The first clause of the quotation describes a motion event that leads to the location of the battle: the king 'advanced' (*saltà dins*), 'forward' (*avant*), to the battleground, and after that we may assume short-range dynamic movements within a limited area of the battleground. The first occurrence of *anar* + Inf, *va-li tal donar* '(he) goes to give him such a blow', allows the motion-then-action reading, but only if we interpret this verbal form as the historical present. In contrast, this reading must be excluded in the case of the following occurrence of *anar* + Inf, or at least it would be strange: after giving a big blow to a man with his sword, the king pulled the sword out of the man's head. These two movements must have taken place one after the other in quick succession, which does not imply any movement in space. After that the king turns to attack another man: in this case it is possible that he took some steps in the battlefield (if we interpret the verbal form as a historical present), although it was not absolutely necessary, because participants in a battle scene are supposed to be close to each other. This scene, if imagined as a dynamic battle scene, does not suppose long movements, although small movements cannot be excluded in two of the three occurrences. However, these contexts are different from those in which the participants travel long distances (i.e. they go to another city), and we are dealing with not only a change of location, but also with a change of scene (as in Example (69)).

3. Finally, in some occurrences the literal interpretation of *anar* does not result in a meaningful reading of the utterance (see occurrences in Sections 3.5.2 and 3.5.3). These occurrences must be regarded as grammaticalized forms, interpretable as referring to completed actions, and they testify that semantic change has already been completed.

The distribution of the kinds of context discussed above is presented in Tables 23–26 in two different ways. First, let us interpret the results of this research as if we accepted the historical present hypothesis, i.e. as if we accepted the motion reading as also possible in the present tense occurrences of *anar*.

Table 23. Distribution of supposed contexts in occurrences of the Catalan *anar* + Inf in the corpus

	Jau	*Desc*	*Munt*	*Per*	*Parl*	*EpC*	
Motion possible	38 (~84.44%)	75 (78.125%)	133 (~56.83%)	1 (~7.14%)	1	-	248
Short-range movement	2 (~4.44%)	7 (~7.29%)	30 (~12.82%)	-	-	-	39
Motion not possible	5 (~11.11%)	14 (~14.58%)	71 (~30.34%)	13 (~92.85%)	-	1	104
Total	45	96	234	14	1	1	391

As discussed above in Section 2.6.7, it seems to be improbable that the forms in question would be conjugated in the historical present. If we reject the historical present hypothesis, we must regard the present tense occurrences as grammaticalized forms, that is, we must assume that the verb *anar* does not convey motion meaning in them. Consequently, all these occurrences must be moved to the group where the motion reading is not possible. The distribution according to the tense of the construction is shown in Table 24.

Table 24. Distribution of supposed contexts in occurrences of the Catalan *anar* + Inf according to the tense of *anar*

		Jau	*Desc*	*Munt*	*Per*	*Parl*	*EpC*	
Motion possible	Present/ preterit syncretism	21	-	6	-	-	-	27
	Preterit	15	60	62	-	1	-	138
	Total	36 (80%)	60 (62.5%)	68 (~29.05%)	-	1	-	165
Short-range movement	Present/ preterit syncretism	2	-	-	-	-	-	2
	Preterit	-	4	4	-	-	-	8
	Total	2 (~4.44%)	4 (~4.16%)	4 (~1.7%)	-	-	-	10
Motion not possible	Present	2	27	157	13	-	1	200
	Present/ preterit syncretism	1	-	-	-	-	-	1
	Preterit	4	5	5	1	-	-	15
	Total	7 (~15.55%)	32 (~33.33%)	162 (~69.23%)	14 (100%)	-	1	216
	Total	45	96	234	14	1	1	391

The data in Tables 23 and 24 reveal that, regardless of whether or not we accept the historical present hypothesis, there is a decrease in the number of contexts allowing for the motion reading and a spread of contexts disallowing it. Nevertheless, this tendency is more obvious if we reject the historical present hypothesis.

A similar correlation can be examined between context types and the tense of the finite verb. The tables below show the possible readings according to the tense of *anar*. In Table 25 it is worth observing that in the case of the preterit examples there is a reduction in the number of contexts where the motion reading is impossible. It may be supposed that the reason for this reduction was that present tense occurrences were used to convey such readings. Table 26 reflects the fact that present tense occurrences are associated with context types that dis-

Table 25. Distribution of supposed contexts in occurrences of the construction *anar* + Inf in preterit

	Jau	Desc	Munt	Per	Parl	
Motion possible	15 (~78.94%)	60 (~86.95%)	62 (~87.32%)	-	1	138
Short-range movement	-	4 (~5.79%)	4 (~5.63%)	-	-	8
Motion not possible	4 (~21.05%)	5 (~7.24%)	5 (~7.04%)	1	-	15
Total	19	69	71	1	1	161

Table 26. Distribution of supposed contexts in occurrences of the construction *anar* + Inf in present

	Jau	Desc	Munt	Per	EpC	
Motion possible	2	15 (~55.55%)	65 (41.4%)	1 (~7.69%)	-	83
Short-range movement	-	3 (~11.11%)	26 (16.56%)	-	-	29
Motion not possible	-	9 (~33.33%)	66 (42.03%)	12 (~92.3%)	1	88
Total	2	27	157	13	1	200

allow the motion reading. Similarly to the analysis of the data in Tables 23 and 24 above, we can note that these tendencies would be more impressive if we rejected the historical present hypothesis.

2.7 Summary: semantic change in the grammaticalization of the Catalan *anar* + Inf construction

In this chapter we have examined the initial stage of the grammaticalization of the Catalan *anar* + Inf construction from a historical pragmatic perspective. The study has been carried out relying especially on a historical corpus of Catalan and Spanish texts from the 13th to the 16th century. I have also included in the research occurrences of the medieval Spanish *ir* 'go' *(a)* + Inf construction. In their origins, they were both purposive constructions that referred to motion with the aim of carrying out a certain act: 'go in order to do sg'. The study of the morphology of these periphrases has provided the following results. Occurrences of the Catalan *anar* + Inf with the purposive preposition *a* 'to' between the finite verb 'go' and the infinitive are very rare at the beginning, but later their use was extended. The prepositional version seems to have a more restricted use in the medieval language state than the use of the construction without a preposition, and the latter use tends to appear in earlier texts. It may be hypothesized that the original version of the construction under study was without any preposition. The results of my study and those of other authors suggest that the prepositional version may have proliferated in Catalan in the course of time in order to differentiate those cases where the verb *anar* appeared as a main verb, with the meaning 'go'. Therefore, the spread of the prepositional form reflects the ongoing grammaticalization of the construction without a preposition.

As regards the verbal tense in which the verb 'go' is conjugated, the following results have been obtained. Differences in the evolution of the Catalan and the Spanish constructions could already be detected in the medieval language state, as manifested in the distribution of present and perfective past tense versions of the Catalan and the Spanish constructions. The analysis of occurrences in their contexts has revealed that the present tense version of the Catalan GO-construction in a preterit context is a later development, which may have proliferated as a consequence of the ongoing grammaticalization of the construction, and the earliest occurrences may have had the verb *anar* 'go' in the perfective past tense (cf. Juge 2006, 2008). It can be argued that the present tense occurrences do not reflect a historical present usage but can rather be interpreted as grammaticalized forms. The Spanish texts of the corpus have not reflected the same spread of present tense occurrences.

After discussing the morphological characteristics, I examined the semantic change of *anar* + Inf, and I based my analysis on occurrences without a preposition, with the finite verb conjugated in the perfective past. The results concerning semantic change in the grammaticalization of the Catalan perfective past (*anar* + Inf) have painted a different picture from that provided in the previous literature.

In order to test the hypotheses of the previous literature concerning the semantic change of the Catalan *anar* + Inf construction, I included occurrences of the intentional periphrasis *pensar (de)* 'think of' + Inf and the inchoative construction *començar (a/de)* 'begin to' + Inf in medieval Catalan and Spanish.

I presented the types of infinitive which appeared in these three constructions in the medieval language state by analysing entrenched local schemas, on the supposition that the semantic change in the Catalan GO-construction might have emerged from these very frequent uses. I examined the use of infinitives in these periphrases, relying on a detailed contextual analysis, concentrating on similarities and differences. The comparison of the uses of these three periphrases revealed a certain type of functional similarity: all three constructions were used in order to attract the listeners' attention. However, a more detailed qualitative analysis of the contexts made obvious a closer similarity between the uses of *ir (a)/anar* 'go' + Inf and *pensar (de)* 'think of' + Inf, and a difference in both regarding the use of *començar (a/de)* 'begin to' + Inf. The verbs *començar* 'begin', *pensar* 'think', and *ir/anar* 'go' all occurred with similar types of infinitives, but they added different nuances to the description of the event. In the case of the BEGIN TO-construction, the event in the infinitive was described as a durative one, but one which was interrupted as soon as it began. With this strategy, the subsequent wedged-in events were emphasized. In contrast, in the case of the GO- and THINK OF-periphrases, the event described in the infinitive was highlighted as a whole, a perfective event with considerable consequences.

These findings throw a different light on the history of the Catalan GO-construction under study and relate its origins to the salience of the meaning component 'intention', suggesting that Detges's (2004) inchoative aspect hypothesis concerning the origin of the Catalan GO-past is mistaken. The historical analysis has revealed that occurrences of the Catalan *anar* 'go' + Inf refer to completed actions and not to their beginning. GO + Inf in medieval texts describes the action as a whole, and its occurrences are often followed by a description of the consequences. The perfectivity inferable in most contexts of use might have played a decisive role in the semantic development of the Catalan *anar* + Inf construction.

To sum up, I argue that, despite the different endpoints of their developments, the histories of both Romance GO-constructions under study run parallel for a time. Both the Catalan *anar* 'go'+ Inf and the Spanish *ir (a)* 'go (to)' + Inf had a stage in their semantic evolution when the meaning component 'intention' was salient within the semantic structure of the verb 'go'. At this stage the construction expressed the intention of the subject and not an 'abrupt start' (i.e. inchoative aspect).

Regarding the topic of the text, descriptions of battle scenes contain most of the utterances with these GO-periphrases, where they seem to fulfil a function of dynamification, intensifying and drawing the addressees' attention to the event which the speaker considers to be the most important in the narrative sequence. However, taking further criteria into account, various types of context have been differentiated. Within the group of contexts where the verb *anar* occurred in its literal meaning, I have differentiated between two main types. In the first type the motion meaning could be assigned exclusively, while in the second type a semantic ambiguity could be revealed. Semantic ambiguity is one of the most important characteristics of contexts in semantic change, because it triggers pragmatic inference mechanisms. Thus, I have concentrated on these semantically ambiguous contexts, where both the original meaning and another meaning could be assigned. Some of the earliest occurrences already suggest that the construction might have possessed another function beyond conveying motion in space. Those contexts in which – although the literal interpretation of the verb *anar* was possible – expressing motion in space seemed to be redundant are significant too. In other contexts, the discourse would have remained coherent and meaningful with the omission of the verb *anar*, that is, the addressee may have interpreted the verbal construction in another way. These types of contexts may have triggered semantic change.

However, grammaticalization may have been supported by other factors, as well. Occurrences with a possible literal meaning could be grouped in another way. In some contexts, the utterance containing the occurrence of the GO-construction continued in a way that has allowed us to conclude that the goal of the motion expressed in the infinitive was thoroughly performed, while in other contexts this conclusion is not possible. In the former contexts, although the perfectivity of the action was not explicitly expressed, it has had to be inferred in order to have a coherent discourse. The frequent use of the construction in this type of context might have established a strong associative link between the use of the periphrasis and this inferential content. Finally, another type of context should be mentioned: where the motion described by the verb *anar* was presumably performed in a very narrow space. Moreover, some of these contexts gave rise to the question of whether it was reasonable to suppose the source meaning at all. In those cases, the discourse remained meaningful without supposing any motion. These contexts differ in this sense from those in which the motion covers long distances and leads to a different scene.

The analysis of the distribution of context types provided the following results. The number of contexts that allow the motion-then-action reading has been decreasing over time, while a spread of context types in which this reading is impossible has been detected. In the case of preterit occurrences, a decline has been revealed in the use of the periphrasis in contexts disallowing the literal interpretation of *anar*. At the same time, the present tense occurrences of the construction seemed to be linked to contexts in which the source meaning is impossible.

Methodological part of the study: data and argumentation in historical pragmatics

3.1 Introduction: methodological problems in historical pragmatics research

Pragmatically oriented grammaticalization research requires a major degree of methodological awareness because of the higher levels and more inaccessible aspects of meaning in which we are interested. There is no doubt that, for historical research which adopts a pragmatic point of view, the data problem is especially relevant, since historical linguistics and pragmatics both have their own methodological difficulties: in a nutshell, how to obtain sufficient and sufficiently reliable data. Historical pragmatics is affected by the data problem in multiple ways, so it is not surprising that the data problem has been emphasized in this area from the beginning, when Jucker (1994) and Jacobs and Jucker (1995) considered the feasibility of this sub-discipline in their pioneering papers. Fitzmaurice and Taavitsainen (2007) also regard the future of historical pragmatics as determined by methodological issues. They believe that

> [A] step forward can be taken by raising the level of awareness of methodological problems. This can be done by explicit discussions on what and how methods are applied, what their limitations are, and what can be done in the future. We can learn more about data, be more systematic in contextual analysis, ask questions about the reliability of our sources and consider their special restrictions.
>
> (Fitzmaurice and Taavitsainen 2007: 29)

Issues concerning data, evidence, methods, and argumentation in linguistic studies are central topics in current metalinguistic discussion as well. The data problem in linguistics in general has been addressed in Lehmann (2004), Kepser and Reis (2005), Penke and Rosenbach (2004/2007), and Kertész and Rákosi (2008, 2012, 2014a and 2014b) among others. In historical linguistics and pragmatics, Fischer (2004, 2007), Lass (1997), and Jucker (2009), and in historical pragmatics, Jucker (1994), Jacobs and Jucker (1995), Fitzmaurice and Taavitsainen (2007), and Navarro (2008) have discussed problems concerning data use.

In this volume I investigate the data problem primarily in historical pragmatics. This area of linguistics is affected by the data problem in a double way. On the one side, due to the historical character of the research, we must sometimes work with a more restricted set of data sources, and, as a consequence, the empirical support for historical linguistic hypotheses is more problematic, particularly as regards spoken language. The diachronic nature of the research presupposes a major scarcity of data because it lacks some data types accessible in other areas of linguistics. Historical pragmatics only has access to a subset of the data used in synchronic pragmatics. Németh T. (2006: 120) indicates three methods of data collection which must be applied together in pragmatics research, namely, intuition, corpus, and experiment. She adds discourse completion tests as a fourth method. However, in most historical research projects, only the first two are available; data from the third and fourth methods are accessible only indirectly. Jacobs and Jucker (1995: 7) have already pointed out that the circumstances of earlier utterances can never be reconstructed in their entirety, so we can only ever obtain an incomplete picture of language change. The historical pragmatician can only rely on written sources as 'approximate evidence'. On the other side, in pragmatics we are dealing with linguistic phenomena which are difficult to access, such as implicatures and other inferences. The methodological problems raised by the nature of the research are also present in historical pragmatics, since these difficult to access aspects of meaning are also examined in pragmatically oriented historical studies, since they can play a crucial role in linguistic change.

This chapter, which deals with methodology, is structured in the following way. Following this introduction, Section 3.2 summarizes data sources used in pragmatics, and particularly in historical pragmatics, while Section 3.3 overviews these sources in research practice. Contextual analysis, which plays a central role in historical pragmatics methodology, is discussed in Section 3.4, while Section 3.5 deals with methodological problems concerning how to assign meanings to historical utterances. Section 3.6, after presenting methodological considerations concerning the notion of data, provides an overview of argumentation methods in historical pragmatics. Through an example taken from the process of research presented in Chapter 2, it also discusses how we can decide between competing hypotheses.

The research questions dealt with in the methodological section of this book are the following:[60]

i.   What data and methods are accessible in historical pragmatics for studying semantic change in grammaticalization and its cognitive background?

ii. What role does context play in semantic change in grammaticalization and in historical pragmatics methodology?

iii. What methods and principles are worth following when assigning meanings in historical research?

iv. What would be the proper notion of data in historical research?

v.   What argumentation techniques are used in historical pragmatics?

vi. How can metatheoretical considerations help us to decide between rival hypotheses during the research?

In the following two sections I aim to answer the methodological research question (i).

3.2 Data sources in pragmatics and in historical pragmatics

Jucker (2009: 1615–1619) discusses the 'armchair', 'field', and 'laboratory' approaches as three different ways of doing pragmatics research. However, the three types of data[61] these approaches provide, namely, data based on introspection or the researcher's intuition, naturally occurring data, and elicited data, can be regarded as the three basic data types in every kind of linguistic research.[62] Let us consider these three types of data on the basis of Jucker (2009).

1. *Data gained by the 'armchair' method.* The source of this type of data can be the researcher's intuition and introspection, but, in addition, via the interview method, opinions and assessments from speakers of a language or language variety can be elicited. Grammaticality judgments also belong to this type.

2. *Data gained by the 'field method'.* According to Jucker (2009: 1615–1618), the field method consists in observing naturally occurring data, that is, utterances that are produced for communicative reasons outside of the research project they are collected for. Jucker mentions different methods as belonging to this approach. With *the notebook method* researchers take notes of occurrences encountered in daily life, with *the philological method* they collect occurrences from fictional material or from written documents, under *the conversation analytical method* transcriptions of actual conversations are used, and, finally, *the corpus method* employs computerized search techniques on electronic corpora.

3. *Data gained by the 'laboratory' method.* The third type mentioned in Jucker (2009: 1618–1619) is the laboratory method, which incorporates different elici-

tation techniques to prompt speakers to produce certain utterances and communicate outside their own intrinsic communicative intentions. The two main types are discourse completion tests and role-plays. In this method, informants 'are asked to imagine communicative situations and to state how they would behave in such situations or how they expect other people to behave in these situations' (Jucker 2009: 1618); consequently, linguistic intuition also plays a role here. However, in contrast to the first type of data, linguistic judgments do not appear here directly; instead, this data type relies on utterances expected to be used in imaginary situations. Although this method provides 'artificial' occurrences and communicative situations, it enables the researcher to have control over many different variables.

Let us discuss whether these types of data are accessible in historical pragmatics. Historical research lacks some of these data types. Although researchers' intuitions as a data source play a role also in historical research, in order to judge the grammaticality of linguistic phenomena of earlier language stages researchers have to rely on what has been termed 'substitute competence' (see below under 3.6.2.2). Native language speakers' linguistic intuitions concerning earlier language stages can be accessed only indirectly: through analysing works of grammarians of the period, remarks in contemporary grammars concerning normative language use, differences between original manuscripts and possible later duplicates or translations, remarks in the margins of historical documents, translations etc. The quantity of obtainable information may differ, and moreover, it may be totally lacking.[63] The philological and corpus methods can only be used to investigate language stages from which we have written documents. Other language stages can only be studied by relying on reconstructed data. Methods which rely on the help of native language informants cannot be applied in historical linguistics. Thus, neither the notebook nor conversation analytical methods are available for the investigation of earlier language stages, and, in the absence of native speakers, elicited data cannot be collected, either. Finally, it is worth noting that data sources not available in the investigation of earlier language stages can still play a role in an indirect way, since historical argumentation can use information concerning the current language state.

The central role of occurrences of a certain linguistic unit found in historical documents is unquestionable in historical research. Fischer (2004: 730) holds that 'the historical linguist has only one firm source of knowledge and that is the historical documents'. It is obvious that this is also true for historical pragmatics. The importance of quantitative data in the study of some central phenomena of historical pragmatics is emphasized, among others, by Fitzmaurice and Taavitsainen (2007: 27), who highlight the importance of the quantitative analysis of large databases. I have discussed the data problem in historical pragmatics in several studies (Nagy C. 2008b, 2009, 2010b, 2014), with special focus on the indirectness and theory-dependence of data. An examination of research practice reveals that historical pragmatics should never stop extending its data sources.

According to Fischer (2004; 2007: 15–17), data sources can be broadened if we are sensible to the whole language state we are investigating and do not only consider the linguistic unit which is the proper subject matter of our study, but also consider other linguistic forms and structures as well, which are formally and functionally similar. Similarly, in historical pragmatics we can look at constructions which are not only formally, but also functionally similar. I have also applied this method in the research presented in Chapter 2.

The considerations presented above demonstrate that we must work with a more restricted set of data sources in historical research, and, as a consequence, the empirical support for historical linguistic hypotheses is more problematic, particularly as regards the spoken language. The paucity of data sources can be counterbalanced by comparing text styles (e.g. on the formal–informal continuum), integrating written sources which are based on oral style (political speeches, sermons, homilies, and personal letters etc.), and using metalinguistic sources (etiquette books, grammars, linguistic works, transcriptions of other types of conversations) (see Jacobs and Jucker 1995, as well as Fischer 2007). In addition, we can use typological data and genetic linguistic information, which can rely on internal or comparative reconstruction (Fischer 2007: 44), information concerning the current language state or linguistic diversity, and data originating from the theoretical framework we are working in (concerning this latter type of information, see Heine 2003/2005: 580, 585–586). Finally, Fischer (2007: 44) emphasizes that the inaccessibility of native speakers' competence concerning the language stage under study might be compensated for by relying more deeply on quantitative information.[64]

Finally, historical pragmatics focuses on levels of meaning which are difficult to access, and for which data collection is very difficult. Fitzmaurice and Taavitsainen (2007: 18) underline the difficulties corpus-based methods face in the investigation of several pragmatic phenomena, e.g. discovering conversational implicatures. Due to the central role that analysis of written sources plays in historical pragmatics methodology, according to Fitzmaurice and Taavitsainen (2007: 27) the challenge of historical pragmatics consists in 'how we adapt corpus-linguistic methodology, which focuses on form, for use in historical pragmatics, which focuses on function, and how the context in its various forms should and can be taken into account'. I discuss these problems in more detail in Section 3.4.

3.3 Data in historical pragmatics research practice

3.3.1 Data sources of the research presented in this book

The study discussed in the present volume has been carried out by relying on a wide range of data and methods traditionally accepted in historical pragmatics, and especially on a historical corpus of Catalan and Spanish texts from the 13th to the 16th century (see 1.3.2). I have investigated current uses of the periphrases under study in the present-day language state of Catalan and Spanish (see 2.1.2), and I have included typological data concerning other languages of the world which use a parallel GO-construction and widely attested paths of semantic change of the construction type 'go' + Inf (see Sections 2.1.1 and 4.3). When interpreting historical utterances, I have also relied on encyclopedic information concerning the historical documents in my corpus (e.g. information about the circumstances and data of their composition). As an example, I have considered the two occurrences of 'go' + Inf with the preposition *a* 'to' found in Desclot's Chronicle as belonging to a later language stage, because both occurrences appear in a title, and we know that titles were often added later to historical texts (see 2.5.2). Concerning the early use of the Catalan *anar* + Inf construction, we have some metalinguistic evidence too, which I discuss in the following section relying on relevant literature.

3.3.2 Metalinguistic evidence

Both Colon (1959/1978: 122) and Pérez Saldanya and Hualde (2003: 56, quoting Badia i Margarit 1950: 143) mention a text containing normative prescriptions from the second half of the 15th century, entitled *Regles de esquivar vocables o mots grossers o pagesívols* ('Rules of how to avoid rude or rustic words'), which criticizes the use of certain words in contemporary Catalan and proposes more correct solutions. Rule 49 of this collection of linguistic prescriptions is quoted in Colon (1959/1978: 122):

> Rule 49: [avoid] *Vaig anar* and *vaig venir* for *aní* and *venguí*, and similar forms.[65]

Aní and *venguí* are 1Sg synthetic preterit forms of the verbs *anar* 'go' and *venir* 'come' respectively, while the verbal forms *vaig anar* and *vaig venir* are the corresponding analytic forms. Pérez Saldanya and Hualde (2003: 56), relying on Badia i Margarit (1950: 143), quote the Rule 48 of the same work too:

> Rule 48: These expressions like *vaig anar* do not seem to be good expressions to Hierony Pau or to me, Pere Miquel Carbonell. It is better to say: *anam, venguem*; not: *vam anar*.[66]

Anam and *venguem* are the 1Pl simple past forms of *anar* 'go' and *venir* 'come' respectively, while *vam anar* is the corresponding 1Pl analytic past of the verb *anar*.

In addition, Colon (1959/1978: 122) also quotes from Lluís d'Averçó's medieval grammar (*Torcimany*) from the turn of the 14th and 15th centuries:

> The second manner is when somebody uses the following words as a way of introducing a quotation: 'Aquelh *me va dir* aytals paraulas.'[67] Although it would be enough to say and it would be fine and not so redundant: 'Aquelh *me dix* aytals paraulas',[68] because if you take a good look, the latter means the same as the former, and because of these two added words, that is, *me va*, it becomes redundant, so they are superfluous.[69]

The quotations above suggest that although contemporary grammarians rejected its use, regarding it as vulgar, this Catalan construction originated in a spoken-language narrative technique and was already used as a real preterit tense in that period. Colon (1976/1978: 146) also refers to metalinguistic evidence concerning French, where the analogous construction was in use even at the beginning of the 17th century. According to Colon, Oudin writes in his grammar of French from 1640: '*Il luy va dire* is a vulgar expression used in place of *il luy dit*.'[70] As the quotation shows, the form *va dire* was in use in vulgar language in order to express the meaning '(he/she) said/told', while in normative language use the synthetic form *dit* was accepted.[71]

3.4 Notions of context in historical pragmatics

In this section I aim to answer the following research question:

ii.   What role does context play in semantic change in grammaticalization and in historical pragmatics methodology?

3.4.1 The role of context in historical pragmatics

Contextual analysis is the most important tool in historical pragmatics methodology. In order to shed light on changes in semantic structure, we need to conduct a sensitive contextual analysis (cf. Diewald 2002; Heine 2002; Bybee 2003/2005: 607; Fischer 2004). Context has a dual character from the point of view of historical pragmatics. On the one hand, both coded and contextual information are involved in the composition of utterance meaning, and in this sense, context triggers semantic change: the meaning of linguistic constructions changes in, and as a result of, ordinary language use in context. On the other hand, the occurrence or non-occurrence of a certain linguistic construction in a certain context in historical documents can provide data about the process of semantic change. In other words, historical occurrences in certain types of context can provide evidence for a previous semantic change.

It is a widely accepted hypothesis in historical pragmatics that grammaticalization requires specific contexts in order to occur. Historical pragmatics research on changes in semantic structures raises the methodological question of how we can reveal the dynamism of the process. From a methodological point of view, the two aspects of semantic change, that is, the innovation and the spread of the change, are hard to separate. This is due to the fact that, in practice, we have only one method to investigate changes in earlier periods: examining the particular written utterances as they occur in historical texts. First, we have to explore nuances of meaning in historical texts by considering several historical occurrences and then to draw conclusions concerning where they come from. However, in order to reveal semantic nuances, we need a sensitive contextual analysis. Only contextual analysis can shed light on changes in semantic structure. After identifying types of contexts accessible in different language stages, we must reveal ambiguities, because these are contexts which trigger grammaticalization, inviting a new interpretation. The description of changes in the semantic structure is closely linked to the description of contexts in which the grammaticalizing linguistic item is used (cf. Heine's (2002) *switch context*, which he considers to be 'an interaction of context and conceptualization'; see below under 3.4.2.2). To reconstruct changes in semantic structure and the spread of the innovation, we should examine the distribution of different context types among successive language stages.

The process outlined above can be considered a micro-level approach to semantic change. Micro-level accounts concentrate on concrete cases of grammaticalization in order to reveal the semantic evolution of a certain linguistic unit in a certain language (cf. Heine et al. 1991). In contrast, considering semantic change through a macro-level approach, several 'cognitive paths' can be observed along which meanings evolve in the different languages of the world. The macro-level approach relies on data and findings provided in grammaticalization case studies which describe such processes in different languages.

Cross-linguistically frequent tendencies of semantic developments can be explored by checking meanings of a certain lexeme and of its later corresponding forms at different historical periods of several languages. However, a prerequisite for all macro-level research is data from micro-level studies about concrete grammaticalization processes in different languages around the world, and it presupposes a previous micro-analysis of meanings in particular contexts of a given language state. It can be seen that the processes of revealing the innovation and its spread and that of sketching cross-linguistically recurrent tendencies in semantic change are closely linked to each other. Therefore, we should conclude that the investigation of contexts is a prerequisite for any kind of historical pragmatics research.

3.4.2 Contexts of semantic change in grammaticalization

3.4.2.1 Semantic change and contexts in historical pragmatics

Semantic change can be conceived of as a change of contexts in which a certain linguistic construction can be used. Context has a dual character from the point of view of historical pragmatics. On the one hand, we can take into consideration the context which is available to language users at the moment of utterance and helps them to interpret utterances. Lexical and contextual pieces of information interact in the composition of utterance meaning. However, since an exhaustive reconstruction of context is hardly possible, everyday language use is always characterizable by a certain vagueness and ambiguity, which may trigger changes in meaning: the meanings of linguistic constructions change in, and due to, ordinary language use in context. In historical research, we have an even more incomplete picture, because the circumstances of earlier utterances are even harder to reconstruct.

On the other hand, we can conceive of context as a key notion in historical pragmatics methodology. Examining context types can help the researcher to reveal semantic change, inasmuch as the occurrence or non-occurrence of a certain linguistic construction in a certain context type in historical documents can provide data about the process of semantic change. Differences in the distribution of context types in successive language stages can reveal how semantic change processes take place, while historical occurrences in certain types of context can provide evidence of a previous semantic change. In the following, we will consider these two notions of context.

3.4.2.2 Context as a triggering factor in semantic change

The nature of context types in which particular stages of semantic change processes can be detected has been the topic of various papers on grammaticalization (Diewald 2002; Heine 2002). Context is conceived of in Heine's (2002: 87) work as 'the crucial factor in shaping new grammatical meanings'. Some general pathways of meaning change in grammaticalization are widely known and accepted in historical pragmatics. However, the need to investigate the intermediate stages is also recognized. Heine (2002: 83) emphasizes that 'without a more comprehensive understanding of the nature of these intermediate stages, generalizations on the overall process must remain preliminary'.

Heine (2002) provides a scenario of grammaticalization, distinguishing between three different kinds of context. The first stage in Heine's scenario is *bridging context*,[72] which triggers an inferential mechanism to the effect that a new, more plausible interpretation is possible, different from the old, lexical meaning. However, this new meaning can still be cancelled, and the old meaning can also be assigned. A certain linguistic form can be linked to more than one bridging context. The following stage is the *switch context* stage. Switch contexts are incompatible with some salient property of the source meaning, and the new meaning is the only possible interpretation in them, although it is still context-dependent. Switch context therefore functions as a device to filter out the source meaning (Heine 2002: 86). The third stage is conventionalization, when the new meaning no longer needs to be supported by context; moreover, 'conventionalization contexts can contradict or violate the source semantics' (Heine 2002: 85). At this stage, the grammaticalizing item can be used in new contexts which are different from the bridging and switch contexts. Heine discusses the process of conventionalization of implicatures from the perspective of contexts, intending to reveal what types of contexts can be associated with what stage of this process. Cancelability shows whether the meaning is context dependent or already independent of context. Heine focuses particularly on switch contexts because he considers them to be responsible for the process of meaning change in grammaticalization. At this stage, the emergence of a new grammatical meaning is due to an interaction of context and conceptualization. Heine regards switch context as 'a filtering device that rules out the source meaning' (Heine 2002: 86).

There is a certain contradiction in Heine's account: how can switch contexts be responsible for the process of meaning change if the new meaning is the only possible interpretation for them, since they are incompatible with some salient property of the source meaning? In my interpretation, the fact that a linguistic unit is able to appear in a context which contradicts its source meaning suggests a previous change in meaning. And the context responsible for that previous change may be the bridging context. In consequence, when investigating grammaticalization, characteristics of bridging contexts are worthy of more attention, while the presence of switch contexts can indicate the availability of a new interpretation and, as a result, can help us in dating semantic change.

Diewald (2002) also deals with contexts in grammaticalization, and her model is more strongly based on morphological and structural factors than Heine's (2002) model. She also proposes three successive stages in grammaticalization which are associated with three different types of contexts. First, a lexical unit starts to be used in contexts where it has not been used before. These are called *untypical contexts* in Diewald's model, and in them the new meaning may emerge as a conversational implicature.[73] The name indicates that this context is not typical in regard to any of the possible interpretations, i.e. it does not offer any clue as regards how to interpret the linguistic structure in question, and it does not unequivocally favour either of the prototypical readings. In a given language stage, several untypical contexts may be present together. The second type, the *critical context*, is characterized by multiple semantic and structural ambiguities, and the new meaning presents itself in it as a possible interpretation among others. Critical context triggers inferencing procedures, because 'there is no way to process this structure in a routinized way' (Diewald 2002: 112), that is, conversational implicatures are needed to create a meaningful utterance. The difference between untypical contexts and critical contexts lies in the fact that in an earlier stage of grammaticalization the new structural and semantic possibilities are distributed over different untypical contexts independently of each other, while in a successive stage these semantic and structural factors accumulate in one specific critical context. The rise of the critical context is related to changes which occur in some other location in the linguistic system, and critical context later disappears. Third, the new meaning is isolated as a separate meaning in what is called *isolating contexts* that favour one reading to the exclusion of the other. Isolating contexts help us to identify different meanings as independent from each other. Isolating contexts exist for each competing interpretation in a certain language state, which is an indication of the completed semanticization of these meanings.

The question arises as to whether there is any correspondence between the contexts assumed in the two models. Although the authors reflect on each other's models, it is more complicated to find correspondences between them than it seems at first sight. Heine seems to easily match his contexts to those assumed by Diewald, while Diewald (2002: 117) emphasizes that the assumed stages in both models do not exactly match.

Heine (2002: 84) notes that bridging contexts correspond roughly to what Diewald calls the critical context in her analysis. In contrast, Diewald argues that they should be subsumed partly under the untypical contexts and partly under the critical context. She emphasizes that they cannot be fully identified with critical context, because this latter is defined by semantic and structural ambiguities, while Heine does not take structural factors into account. Furthermore, the critical context disappears at a later stage, unlike contexts assumed in Heine's scenario. As the analyses of examples below reveal, I agree with Diewald's opinion, although I do not rely on her arguments. The fact that Heine's model concentrates on semantic changes and he does not consider structural factors does not mean that, for example, the bridging context is not characterized by structural ambiguity. Actually, the notion of bridging context allows for such an interpretation. A later disappearance is also compatible with the notion of bridging context. However, it is not necessary, while Diewald's critical context obligatorily disappears at a later stage of grammaticalization. Thus, a possible correspondence between bridging and critical contexts depends on the later disappearance or retention of the bridging context.

Heine (2002: 85) relates switch contexts to what Diewald calls isolating contexts, while emphasizing that there can be several isolating contexts that are in semantic and structural opposition. Furthermore, the development of isolating contexts is also due to changes in the grammaticalizing items. In summary, while Heine assumes a closer parallelism between both analyses, Diewald (2002: 117) maintains that 'though principally compatible, both models focus on different aspects of grammaticalization processes'.

After a consideration of the examples provided by both authors, I find Diewald's comparison more valid, as I illustrate below through the analysis of several examples taken from my own research. Finally, it is worth noting that Heine's model, being more general, is easier to apply when analysing actual grammaticalization processes. Diewald's model, due to its emphasis on morphological and structural aspects, gives rise to several theoretical questions. First, whether critical contexts disappear in every process of grammaticalization, and second, whether there is necessarily more than one untypical context in every process of grammaticalization. In grammaticalization processes where there was only one untypical context, it would not be reasonable to differentiate between untypical and critical contexts. All of these are only theoretical considerations, which could be answered only by relying on various case studies.

In the following I will provide some examples to illustrate the context types discussed above.

Heine (2002)

1. As explained above, Heine's bridging context triggers an inferential mechanism and, as a consequence, rather than the source meaning, a new, more plausible interpretation emerges. The target meaning is still cancellable, that is, an interpretation in terms of the source meaning is still also possible. Consider Examples (73) (presented above in Section 2.6.8 as (70)) and (74).

(73)

E con foren al peu de l'escala, madona la reina e sa germana **anaren-se**

                                          go.3Pl.PerfP-Refl

abraçar;   *e estegren així abraçades, e besant e plorant,*

embrace.Inf

'And when they came down the ladder, **at the foot of it**, my Lady the Queen and her sister **embraced (= went to embrace)** each other and remained thus embracing and kissing each other and weeping much'

                                   (Munt I 181,31–33)

(74)

E sobre açò lo senyor rei N'Anfòs pres per la man lo senyor infant En Pere, qui li seïa
de prop, un poc pus baix, e **anà'l**　　　　　　**besar,**　*e dix-li:*
　　　　　　　　　　　go.3Sg.PerfP-Clit　kiss.Inf

'And upon this the Lord King took the hand of the Lord Infante En Pedro who was near
him, but a little lower, and **kissed him (= went to kiss him)** and said to him:'

(Munt II 14,1–3)

The combination of the motion verb *anar* with the infinitives *abraçar* 'to embrace'
and *besar* 'to kiss' in most of the cases is also interpretable literally: 'she/he went
to embrace/to kiss her/him'. However, in Examples (73) and (74) these combinations appear in a context where the spatial position of the participants, and their
spatial proximity renders such a reading implausible. In Example (73) both the
queen and her sister were at the foot of the ladder, i.e. they must have been very
close to each other. The motion meaning cannot, however, be fully excluded, since
they probably took a few steps towards each other. In (74) the context makes the
motion reading of *anar* even more implausible, since after the king had taken the
Infante's hand, he must have been close enough to kiss him. However, we cannot
fully exclude spatial motion, since taking somebody's hand is something that can
be done when standing further away than when kissing. Anyway, the formulation
'he took his hand and then he went to kiss him' sounds strange. So, in both utterances the hearer reasonably assumes that the speaker may have another aim when
using the construction, for instance, to put emphasis on the dynamism of the act.

2. The switch context, as we have already seen, is in conflict with some salient
property of the source meaning, thus the only possible interpretation is a new
reading. Heine puts special focus on this context type, since he maintains that
this is the one responsible for semantic change in grammaticalization. He characterizes this stage as an interaction between context and conceptualization.

(75)

E **anà'ns**　　　　　　　　*abraçar*
　Aux(go).3Sg.PerfP-Clit.1Pl　embrace.Inf

'Then he **embraced (= ?went (to us) to embrace)** us.'[74]

(Jau 224)

In (75) we can observe a conflict due to a salient property of the source meaning:
the verb *anar* 'go' is a deictic verb, which describes motion *away from* the deictic
centre. However, in (75) we are dealing with a motion *to(ward)* the deictic centre,

due to the use of the 1Pl clitic ('us'). This contradiction compels the hearer to find a new interpretation different from the source meaning.

3. At the conventionalization stage, the new meaning no longer needs to be supported by context. Once the target meaning has been conventionalized, it can appear in new contexts, other than bridging and switch contexts.

(76)

Mas con foren prop d'Agda, noves los **van**   **venir**   *con havia*

                          Aux.3Pl.Pres come.Inf

pres, lo dia passat, a aquells de Besers;

'But when they came near Agda news **arrived** (= **?go [in order] to come**) that they of Besers had been taken on the previous day.'

                                    (Munt II 10,36–37)

The combination of *anar* with the infinitive *venir* 'to come', as in (76), is itself a contradiction of the source semantics of 'go', since the source meaning would yield the strange and nonsensical reading ?'the news come/are coming (in order) to go'. The occurrence of the construction 'go' + Inf with an inanimate subject (*noves* 'news') also suggests a more advanced stage of grammaticalization, which is incompatible with the purposefulness present in the original meaning of the purposive motion verb construction. We do not need the context anymore to find out that this is not the literal meaning, i.e. this example is representative of the conventionalization stage. Similarly, in the occurrence in (77) we cannot assume the intentionality of the subject.

(77)

E a la fin lo senyor rei donà tal de la maça al cap del cavall del príncep, que el cavall

fo fora de tot son seny e **va**   **caure** *en terra;*

                         Aux(go).3Sg.Pres fall.Inf

'But in the end the Lord King gave such a blow of his mace on the head of the Prince's horse that he made it senseless and it **fell** (= **?goes to fall**) to the ground.'

                                    (Munt II 56,15–17)

In (77) the infinitive *caure* 'to fall' contradicts the intentionality present in the source semantics of *anar* used in the purposive construction *anar* + Inf. It serves as a good example of conventionalization, since the interpretation of the event as completed is independent of context. The same can be said of occurrences with a patient subject. If we look at examples (73)–(77), we can observe that occurrences of *anar* + Inf with the verb *anar* conjugated in the present tense appear in context types characteristically belonging to the later stages of grammaticalization. In other words, the present tense conjugation presumably implies the conventionalization stage. This supposition, however, contradicts the 'historical present' hypothesis (cf. Section 2.3.2).

Diewald (2002)

1. According to Diewald's model, the new meaning arises in untypical contexts as a conversational implicature. This context does not clearly favour one of the prototypical readings; in other words, both the original and a new reading are accessible.

(78)

E quant fo defora, puyí lo cavayl dels esperons e **anà**   *ferir*   *entre·ls*

                  go.3Sg.PerfP   attack.Inf

sarraÿns, sí que·l primer colp n'abaté, ab lo pits del caval seu, ·IIII· a terra.

'And when he was outside, he dug his spurs into his horse and **went to attack/attacked** the Saracens, so that at first he threw four of them to the ground with the breast of his horse.'

(Desc III 93,23–26)

In the case of the occurrence in (78) (presented above as (27)) both the original meaning ('motion') and a new meaning ('completed event/preterit') can be attributed. This is due to the fact that the act of attacking needs some motion in space. At the same time, it is not explicitly expressed that the act was actually carried into effect, but this can be concluded from the fact that the subsequent clause describes its consequences. The occurrence in (78) would also be a good example of Heine's bridging context.

2. The second type, called the critical context, is characterized by semantic and structural ambiguities, and the new meaning appears in it as one of the possible interpretations. The critical context triggers inferential mechanisms, because the recipient has to resort to conversational implicatures in order to process the linguistic unit and to attribute an adequate meaning to it.

(79)

e tirà la primera pedra lo maestre del fenèvol e errà la brigola; e nós

anam   *pendre*   *lo fenèvol e tiram e donam tal en aquela brigola*

go.1Pl.Pres/PerfP   take.Inf

que la caxa li obrim;

'The master of the "fonevol" shot the first stone and missed the "brigola".[75] We **went/go to take** charge of it, shot, and hit the "brigola" so hard that its box was broken;'

(Jau 462)

In (79) the chronicler describes a battle scene where the warriors try to shoot at the enemy's war-engine with a 'fonevol'. The construction *anam pendre* is structurally ambiguous, since *anam* is a syncretic form, able to express either the present or the past tense in Old Catalan. So, this context allows for multiple interpretations: (a) 'we go in order to take the *fonevol*', (b) 'we went in order to take the *fonevol*', and (c) 'we took the *fonevol*'. In this context, all three readings are possible. The master of the fonevol shoots the first stone and the king the second. In consequence, when the first stone was fired, the king cannot have been right next to the fonevol, and so some motion was necessary. There is another possible reading, according to which the king was close during the first shot, and thus the emphasis is on the dynamism of the act described in the infinitive. This reading is favoured by the location of the utterance in the discourse: the storyteller describes the most interesting event, when the enemy's war engine was broken. The event described in the infinitive was a completed act, as can be concluded from contextual information, since in order to shoot the stone, the fonevol must have been used. The syncretism of the verbal form *anam* leads to ambiguity, since if we interpret it as a present form, the literal meaning is relegated to the background. The utterance in (79) also could exemplify Heine's bridging context. In conclusion, in this process of grammaticalization the bridging context can be identified with Diewald's critical context, since the structural property which may have triggered the change, namely, the present–past syncretism of 1Pl forms, later disappears.

A thorough analysis of occurrences in critical contexts is very useful in examining what interpretations of a construction are possible in a certain language stage, and whether they are context dependent or coded meanings. Diewald's (2002) method also aims to answer this question. Examining the development of German modal verbs, Diewald (2002) analyses the interpretation of what she considers a highly ambiguous construction, because if the wider linguistic context is disregarded, it can be interpreted in at least three ways. How does Diewald know that these three readings are actually accessible at the language state under study? She argues that 'this can be concluded from the fact that there are instances of the critical context in Middle High German where, due to contextual disambiguation, only one of the three alternative interpretations is possible, while the other two are excluded' (Diewald 2002: 111). She also maintains that if there are disambiguating contexts for various readings, it is justified to regard the construction as an ambiguous structure in the sense defined, with these possible interpretations. The wider context, which provides further information, disambiguates the construction.

But how can we determine which is the default (or preferred) reading? Default interpretations are those that we assign for lack of a wider context, i.e. further contextual clues (cf. Levinson 2000: 11). In order to determine this, Diewald aims to reconstruct chains of pragmatic inferences which result in a certain reading. She considers the preferred reading as one which is accessible through more than one chain of inferences, each with different starting points. In the case of the grammaticalization process presented in this book, this preferred meaning is perfectivity (e.g. in the case of the utterance in (79) we can access the perfectivity reading either by starting from the meaning in (a), or the meaning in (b)). Diewald's method highlights the fact that it is not enough to consider only one context; it is essential that several contexts in which the construction under investigation can occur at a certain language state are taken into account together.

3. The third step in Diewald's (2002) scenario is when the new meaning is isolated as a separate meaning in what is called 'isolating contexts' that favour one reading to the exclusion of others. Each reading accessible in a certain language stage has its own isolating context, which shows that the semanticization of the new reading has been completed.

(80)

*Ab tant lo senescalc **se**   **va**   a eyl   **acostar**, e saludà-lo*

          Refl   go.3Sg.Pres   go.Inf

'Upon this the High Seneschal **went (= ?goes to go) up to** him and saluted him.'

(Desc III 160,9–10)

(81)

*E quant venc l'endemà matí, que foren levats,   **anaren**   **oyr**   les misses e*

          go.3Pl.PerfP   hear.Inf

puys tornaren dinar;

'And the next day, when morning came and they get up, they **went to attend a mass**, and then they returned to dine;'

(Desc II 136,12–13)

Utterances in (80) and (81) exemplify isolating contexts of two separated meanings of *anar* + Inf. In (80) we can observe the isolating context of the meaning 'perfectivity/preterit tense', where the original, literal meaning can be excluded on the basis of the meaning of the infinitive. The example in (81) illustrates the isolating context of the literal reading ('motion with the aim of doing sg'). The subsequent clause containing the verbal form *tornaren* '(they) returned' makes it obvious that it refers to an actual motion in space. The accessibility of both readings at the same language stage (occurrences in (80) and (81) are taken from the same text intentionally) shows that the construction *anar* + Inf had already become an ambiguous structure by that period. Later, the two readings were differentiated formally as well, through the insertion of the preposition *a* between the two verbal elements of the construction. These changes have resulted in the current situation: the construction without a preposition conveys the preterit tense meaning (*anar* + Inf), while the preposition is obligatory in purposive constructions where the verb *anar* conveys its literal meaning 'go' (*anar a* + Inf).

The isolating context in (80) would also be appropriate to exemplify Heine's conventionalization context. The isolating context in (81) separates the original/ source meaning. This reading disappears in the later development of the construction, and another construction is now used to convey it (*anar a* + Inf). However, it would be possible that in a given language these readings would not be differentiated formally, and, as a consequence, the construction would remain ambiguous in certain contexts (which is the case with the construction *ir a* + Inf in current Spanish; see Section 2.1.2.3).

Two points should be underlined regarding the issues presented above. First, in order to investigate semantic change, we should be able to assign meanings to historical occurrences and to determine what kind of context we have in each case. For example, Heine's switch context is incompatible with some salient property of the source meaning, and the conventionalization context contradicts or violates the source semantics. However, in order to reveal such incompatibilities and contradictions, we have to know what meaning the lexical items convey in that context. That is to say, in every case the analysis of semantic change presupposes the researcher's knowledge of the meaning of the particular linguistic unit in the historical context. We need to check (a) whether the new interpretation is possible or not, (b) whether it is the only possible reading, or only one of the accessible interpretations, and (c) whether the new meaning is context dependent (i.e. emerges as a conversational implicature) or already independent of context. We should minimize the subjectivity in guessing meanings and base our analysis on linguistic clues whenever possible (see below in Section 3.5).

Second, we can distinguish between two main types of contexts. The first group comprises Heine's bridging context and Diewald's untypical and critical contexts. These contexts trigger grammaticalization, because the old meaning is only one of the possible interpretations in them and they allow for a new meaning as well. Ambiguity is always a necessary stage of a change in meaning. The other type includes Heine's switch and conventionalization contexts and some of Diewald's isolating contexts, namely those that favour the new reading and exclude the source meaning. These are very revealing from a methodological point of view, because they can serve as evidence of a previous change in meaning. In other words, their existence presupposes a previous reorganization of the semantic structure of the grammaticalizing lexical item. To put it differently, although both Heine and Diewald consider context as a triggering device in grammaticalization, there are types of context which provide a methodological clue as well: if we find an occurrence in one of these types of context (Heine's switch and conventionalization contexts and Diewald's isolating context of the new meaning), we can come to a conclusion about which stage of development the grammaticalizing unit may have reached. We have to base our analysis on contexts which favour or disregard one meaning as opposed to another. This second type of context leads us to a consideration of context as a methodological notion.

3.4.2.3 The methodological notion of context

Only a deep analysis of contexts can reveal the particular steps in meaning change processes. But what do we mean by contextual analysis? Context can be conceived of in several ways. Let us examine the definition of 'critical context' in Diewald (2002). As I have mentioned above, Diewald's critical context is characterized by multiple semantic and structural ambiguities, and the new meaning presents itself within them as one of the possible interpretations. But how do we know what reading has to be preferred in critical contexts? Diewald characterizes the attribution of meanings in the following way: 'A reading … is excluded not because the critical context would not allow it, but because of *further contextual factors*' (Diewald 2002: 111, emphasis added). This choice of phrase indicates that Diewald believes critical context only includes a narrow passage, perhaps the clause which contains the morphological construction itself, but 'further contextual factors', which presumably are clues capable of disambiguating the meaning, do not belong to it. Consequently, Diewald's wording implies two types of contexts: a narrow context ('critical context') and a wider one ('further contextual factors'). These latter are presumably clues which disambiguate the meaning. Differentiating between these two context types is significant from a methodological point of view as well, although they are notions of contexts which can also be used to describe linguistic processing. In the following, I discuss primarily methodological notions of context.

Another narrow conception of context appears when we examine what types of verbs, infinitives, subjects, etc. can occur in a certain construction (cf. Bybee 2003/2005). In the grammaticalization process through which new, abstract meanings evolve, the grammaticalizing unit gradually extends to more and more types of grammatical environments. The final stage of this process occurs, for instance, when all types of verbs or subjects can enter a construction without any grammatical constraints. There is no doubt about the importance of considering the combinational possibilities of grammaticalizing items, since quantitative data concerning these aspects of the process inform us about the course of semantic development (cf. Bybee 2003/2005, 2005). It is reasonable to apply frequency counting methods when working with a large corpus where it is not possible to check the larger context of each example. However, it is not sufficient to take into account only the narrow context, because it does not provide enough information to determine what meaning a certain lexical unit expresses. Furthermore, the exclusive use of quantitative data is likely to result in a misguided analysis of semantic change in grammaticalization. Consequently, we have to use further data and to examine historical occurrences in their broad contexts. Navarro (2008: 14–16) points out that in historical pragmatics, quantitative analysis should be complemented with qualitative research, the 'quali-quantitative' method being the most appropriate one, with more reliable results. The quantitative method helps us to avoid the problems caused by subjectivity and introspection (although the use of researcher's intuition cannot be excluded in this case, either), and, at the same time, with the qualitative method we can learn about those aspects of meaning which are difficult to access on the basis of quantitative data.

I have discussed above (under 3.4.2.2) Diewald's (2002) method of simultaneously taking into account several contexts in which the construction under investigation can occur at a certain language state. This method implies a further broad conception of context: we consider each example from the broad perspective of the whole language state and consider several occurrences together. Moreover, when they try to determine what inferencing mechanisms may be at work at a certain historical period, Traugott and Dasher (2002/2004) propose to consider an even broader context. Their answer to the question of how we know when an invited inference (IIN = conversational implicature) is starting to be exploited, is worth quoting:

> This must always be a matter of interpretation, on the assumption that IINs are cross-linguistic. Sometimes an IIN may be inferred because the later history by hypothesis requires an earlier stage in which the IIN has operated, but we must always exercise utmost care in projecting such IINs on the textual data. As a working principle, as long as the original coded meaning is accessible, we should assume that the invited inference is just that, a meaning derivable from the semantics in combination with the discourse context. In written records, clear evidence of semanticization of a polysemy typically comes from the appearance of an item in a 'new' context in which the earlier meaning(s) of the item would not make sense. At a later time the older meaning may or may not disappear; if it does, this is further confirmation of the earlier coding of the former pragmatically invited inference.

> (Traugott and Dasher 2002/2004: 44)

Traugott and Dasher's (2004) method requires an even broader context of interpretation for historical occurrences: the context of the later history of a construction can weaken hypotheses about earlier stages of development or make them more plausible. In other words, during linguistic research, we interpret utterances relying on a much broader context than during everyday linguistic interactions. Such an approach compensates for the deficiency that in historical linguistics we do not have the possibility to reconstruct the context to the same degree that was accessible to contemporary language users.

In my research I have distinguished between two types of contexts: narrow and broad. These narrow and broad conceptions of context are not unknown in pragmatic research. Inspired by Relevance Theory, Bibok and Németh T. (2001a) also establish two types of contexts: *immediate context* and *extended context*. By the immediate context of a word they mean the remaining part of the utterance under investigation, that is, the utterance itself in which the lexical item under investigation occurs (Bibok and Németh T. 2001a: 307; cf. also Bibok 2004: 300). The utterance context can be extended with the help of three types of information: (a) with information from the immediately observable physical environment, (b) with information from the preceding discourse, and (c) by adding encyclopedic information (Bibok and Németh T. 2001a: 300; cf. also Bibok and Németh T. 2001b: 16). Determining the utterance borders is not easy in synchronic pragmatics (cf. Németh T. 1994); however, in historical studies it is even more difficult because of the absence of punctuation in some early texts. Therefore, in historical research, a slightly different conception of contexts is necessary. Because of this, I use the terms *narrow context* and *broad context*. In a historical study, a narrow context must be defined grammatically, as a part of the text which corresponds to a sentence or a clause, containing one finite verb, i.e. it can be shorter than an utterance.[76] The narrow context in my case study can show which infinitives can co-occur with the grammaticalizing item in a construction, but it can also contain other lexemes (e.g. nouns or adverbs) which help to assign meaning to the utterance. To sum up, the main difference between Bibok and Németh T.'s (2001a) immediate context and my narrow context is that a narrow context is defined grammatically and can be shorter than an utterance.

The concept of the broad context mainly coincides with that of the extended context in Bibok and Németh T. (2001a). However, a broad context adds information to a narrow context not only from preceding discourse, but also from subsequent utterances. This is possible because the present definitions of contexts are not intended to describe the composition of utterance meaning, but to be used in contextual analysis as a tool of historical pragmatics methodology. Therefore, from the point of view of the content, I define broad context as a coherent passage of the historical text. Its length can vary on different occasions. The broad context may include sentences or clauses that precede or follow the occurrence of the lexical unit under investigation, and also contain information from the physical environment, encyclopedic information and inferences present in the context. The term *context* does not refer exclusively to textual circumstances. The physical context cannot be reconstructed, but we may have encyclopedic information about the circumstances of the production of historical texts and concepts present in the context. Furthermore, taking into account the broad context or the later history of a construction under study, we can infer the presence of certain pragmatic inferences, which form part of the cognitive context.

A further reason to make a distinction between narrow and broad contexts is that narrow context can be examined with electronic data processing of much larger corpora. We can examine, for instance, the co-occurrence of certain linguistic units, searching for given infinitives, clitics, etc. It is worth noting, however, that written documents from before the spread of a standard spelling system may show a great diversity of spelling, using several different orthographic symbols for the same sound or using abbreviations, which presents a difficulty for computer searches. When considering the broader context, we should examine the occurrences one by one. Such a qualitative approach, even if it is very time-consuming, is indispensable in historical pragmatics. Naturally, the most effective approach is a combination of the qualitative and quantitative methods.

The question arises as to what extent the context should be broadened. The consideration of a broadened context can override our preliminary assumptions about the meaning of a certain construction, formed on the basis of its narrow context. In some cases, it already occurs with the inclusion of only one further utterance. In other cases, the interpretation remains the same in the broad context as it was in the narrow one. It is not worth broadening the context when a new, independent passage begins which probably will not affect the interpretation. In other words, this method is based on semantic factors, and the evaluation of particular cases requires a qualitative analysis, partly based on the researcher's subjective judgment. Even if it may seem that the use of linguistic intuition leads to excessive subjectivity, it cannot be completely excluded. The consideration of broad contexts can serve as a basis to determine the most plausible reading of a historical occurrence, and it can even change previous interpretations or influence how we conceptualize the event described in the utterance.

3.5 Methodological issues of assigning meaning in historical research

3.5.1 Linguistic clues in the contexts of historical occurrences

In order to avoid misinterpretations based on our subjective linguistic intuitions and the projections of our knowledge about the present-day language state, we should find linguistic clues that may help us to reveal what shades of meaning historical occurrences convey. In the following I aim to answer research question (iii) concerning methods and principles worth following when assigning meanings in historical research. I present certain linguistic elements which suggest that the verb GO does not retain its lexical meaning of motion, and that its coded meaning is changing or has already changed.[77] In this sense, such elements reveal undergoing grammaticalization processes. Obviously, not each linguistic clue presented below occurs in each context, and some contexts may contain more of them, while others neither of them. When attributing meanings, I follow the principle according to which if nothing in the context contradicts it, we should attribute the original lexical meaning (cf. Traugott and Dasher 2002/2004: 44; Section 3.4.2.3). If the lexical meaning of the finite verb must be excluded due to a linguistic clue, we should find a new reading suitable in that context.

3.5.2 Linguistic clues in the narrow context

3.5.2.1 Lexical items

Linguistic clues which serve as points of orientation when guessing meanings and thus identifying the degree of grammaticalization may already occur in the narrow context. Sometimes the infinitive itself, and the nouns which appear near it, are semantically irreconcilable with the source meaning of the grammaticalizing construction. In these cases, certain pieces of encyclopedic information are activated that influence the interpretation and exclude the source meaning. Consider the Catalan Example (82):

(82)

va   *avallar*   *del cavall*

go.3Sg.Pres   dismount.Inf

'(Boxadors) **dismounted** (= **?goes to dismount**) from his horse'

(Munt II 176,29)

Example (82) literally means 'he goes to dismount from the horse'. However, we should realize that the lexical meaning 'go' of *anar* has already disappeared here, and that the collocation cannot retain its literal meaning *va avallar* 'goes to dismount', because one does not need to go anywhere to dismount from a horse.[78]

Adverbs that can be semantically linked to the event in the infinitive can also be useful linguistic clues. These adverbs foreground the semantic content of the infinitive, instead of that of the finite verb. Consider the Catalan Example (83):

(83)

e los altres **van**   *ferir*   *en la llur peonada* <u>*tan fortment*</u> *que ab los*

             go.3Pl.Pres attack.Inf   so vigorously

dards ne mès cascun un en terra

'and the others **attacked** the foot soldiers <u>so vigorously</u> that each one threw down one with his dart' / ?'and the others **go** <u>so vigorously</u> **to attack** the foot soldiers that…'

(Munt II 177,18–20)

In Example (83) the construction *van ferir* literally means '(they) are going/go to attack'. However, the adverb *fortment* 'vigorously, hard' likely describes the method of attacking and not the method of going, and thus it foregrounds the meaning of the infinitive. This linguistic judgment is based on a kind of semantic compatibility: normally, an attacking event can be conceptualized as being realized vigorously more easily than would be the case with a going event; i.e. it is a more typical phrase, which describes a more typical situation. The use of such adverbs draws attention to the event described in the infinitive, triggering the *semantic bleaching* of the main verb, whose semantic content moves to the background.

3.5.2.2 Grammatical person

There are some very interesting examples where the goal of the motion event is related to the first grammatical person. In Catalan, when the deictic centre corresponds to the speaker, the verb *venir* is used (cf. '(she/he) came to me/us' vs. ?'(she/he) went to me/us'). Let us consider the Catalan examples in (84) and (85) (presented above in Section 3.4.2.2 as (75), repeated here for convenience):

(84)

E anà's   *gitar*   <u>*als nostres peus* e *besà'ls-nos* e *pregà'ns*</u>
   Aux.3Sg.PerfP-Refl throw.Inf

per Déu que li perdonàssem.

'Then he **threw** (= **?went to throw**) himself <u>at our feet</u>, kissed them, and prayed us in God's name to forgive him.'

(Jau 520)

(85)

E anà'ns   *abraçar*
   Aux.3Sg.PerfP-Clit.1Pl embrace.Inf

'Then he **embraced** (= **?went to embrace**) us'

(Jau 224)

These examples are also presented in Bruguera (1981). Bruguera maintains that the correct interpretation of occurrences in the Chronicle of James I is the literal one, in other words, they are not examples of the periphrastic past tense. He opines there are only three occurrences of *anar* + Inf that allow a preterit reading (Bruguera 1981: 41), and two of them are the occurrences presented here as (84) and (85). However, Bruguera does not explain why he interprets them in this way.

3.5.2.3 Inanimate subject

An advanced stage of grammaticalization may be identified when the originally purposive construction 'go' + Inf co-occurs with a subject that is inappropriate to the source meaning. The motion-with-intention meaning excludes inanimate subjects, thus occurrences such as (86) (presented in 3.4.2.2 as (76), repeated here for convenience) may suggest that semantic change has already occurred:

(86)

Mas con foren prop d'Agda, noves los **van**   **venir**   *con havia pres, lo*
                                                   Aux.3Pl.Pres   come.Inf

dia passat, a aquells de Besers

'But when they came near Agda news **arrived (= ?go [in order] to come)** that they of Besers had been taken on the previous day.'

(Munt II 10,36–37)

3.5.2.4 Subject with the thematic role of patient

A similar incompatibility can be observed in occurrences with animate subjects which, however, have the thematic role of patients, as in (87) (presented above in 3.4.2.2 as (77) and repeated here for convenience):

(87)

E a la fin lo senyor rei donà tal de la maça al cap del cavall del príncep, que el cavall

fo fora de tot son seny e **va**   **caure**   *en terra;*
                                       Aux.3Sg.Pres   fall.Inf

'But in the end the Lord King gave such a blow of his mace on the head of the Prince's horse that he made it senseless and it **fell (= ? goes to fall)** to the ground.'

(Munt II 56,15–17)

3.5.3 Linguistic clues in the broad context

In other cases, we have to broaden the context in order to be able to find the most plausible reading of a historical occurrence. The content of the preceding or following clauses or sentences can make it unlikely the literal meaning will be assigned. For instance, if the sentence that contains the construction in question is followed by another sentence that describes the consequences of the event, we can suppose that the periphrasis represents a completed action and not only the motion with the aim of carrying out that action. Consider Example (88):

(88)

venc a un portell altre, e **anà**   **ferir**   *lo cavall per la testera,*
                                       go.3Sg.PerfP   wound.Inf

e <u>el cavall estec estabornit</u>

'(she) aimed at another opening, and **wounded (= went to wound)** the horse in the head, and <u>the horse was stunned</u>'

(Munt I 198,6–7)

In the Catalan Example (88), the construction *anà ferir* must convey the meaning 'wounded', because the following clause describes the consequence of the action of wounding. If we interpreted *anà ferir* as 'went to wound', the information that the horse was stunned would seem to be the consequence of the motion and not of the attack, which would not make sense; alternatively, the motion in the first clause and the stunning in the second one would appear as two events independent of each other, which would break the coherence of the story. In order to maintain the coherence of the story, we have to suppose the following events: (a) he went to wound the horse, (b) he wounded it, and (c) the horse was stunned. The supposition that the periphrasis *anà ferir* conveys a perfective reading and is not used to describe a motion in space is also confirmed by the content of the previous clause: it describes a motion leading to the location of the act of wounding. This utterance reveals the presence of an inferencing procedure. However, we do not know whether the information that the action described in the infinitive was actually performed is present as an inference or forms part of the coded meaning of the periphrasis. In any case, it must be present somewhere in the context: the event in the infinitive is foregrounded, and if there is any reference to motion at all, it remains in the background. The Catalan Example (89) is similar to (88), but now the following clause describes an event which is not a consequence of the action in the infinitive, but presupposes its performance.

(89)

Mas los almugàvers **anaren**   **trencar**   *les lisses e el mur de la vila, e*
                     go.3Pl.PerfP break.Inf

<u>*entraren dins*</u> *e los cavalers ab eyls*

'But the Almogavars[79] **went to break/broke** the entrenchment and the wall of the city, and <u>entered</u> and the knights with them as well'

(Desc III 146,13–16)

Like in the previous example, the information that the action described in the infinitive was actually performed seems to be present in the context of (89). We have the information from the broad context that the Almogavars entered, and we can therefore infer that *anaren trencar* 'they went to break' does not refer to the motion with the aim of breaking the wall of the city but to the breaking itself, because in the Middle Ages this act was necessary for the enemy to enter into the city. Although we know that the breaking of the city wall was actually performed, we cannot be sure that this information comes from the coded meaning of the periphrasis. It may also be present only as an inference in this context.

The broad context may also contain an adverb clause of place or time that makes the motion meaning implausible. Consider the Catalan occurrence (90) (a fragment of (70), presented in 2.6.8), where the content of the clause that precedes the one containing the periphrasis refers to a location and renders the motion meaning implausible.

(90)

E <u>con foren al peu de l'escala</u>, madona la reina e sa germana **anaren-se**

go.3Pl.PerfP-Refl

abraçar

embrace.Inf

'And when they came down <u>the ladder</u>, <u>at the foot of it</u>, my Lady the Queen and her sister **embraced/(went to embrace)** each other'

(Munt I 181,31–33)

In Example (90) the first clause provides the information that the participants were already at the foot of the ladder, where the event described in the infinitive took place; that is, they presumably did not have to go anywhere. The very action of embracing each other seems to be emphasized by the use of the periphrasis, and the motion meaning seems to be irrelevant. Some motion seems to be possible, but it is likely a minimal movement in space. By the use of the GO-periphrasis, the speaker highlights the non-ordinary nature of the performance of the action in the infinitive, by referring to the intentional world of the subjects. This is possible because 'intention' and 'dynamism' are inherent components of the meaning 'go' in this purposive construction.

Subsequent clauses may also contain information excluding the motion meaning, as can be observed in (91).

(91)

e va   trer   lo bordó que portava cint

   go.3Sg.Pres   take.Inf

'and he **took** (= **?goes to take**) the bordon[80] <u>he was wearing in a belt</u>'

(Munt II 176,25–26)

In Example (91), *va trer* literally means '(he) is going/goes in order to take'. The infinitive *trer* 'to take' in itself does not exclude the motion meaning of *va*: it is the content of the following clause which does not allow it. The meaning ?'he goes to take the bordon he wore in a belt' cannot be assigned because the thing the agent took was something on his own body. In descriptions of battle scenes, several objects can be mentioned whose localization is not explicit, but they are presumably on the subject's body or close beside it, or leastways we assume so relying on our word knowledge (e.g. various pieces of armour or clothing) (cf. also the example in Chapter 3 n. 31, in a present tense context). Our word knowledge influences the interpretation of the construction as well. In other words, the utterance in (91) cannot mean that 'he goes to himself to take his bordon'; instead, we assign the meaning of a completed action or a past tense value to the construction. This reading is confirmed by the verbal tense of the verb in the subsequent clause. The verbal form *portava* 'was wearing' is in the imperfect past tense. It is a descriptive past tense, which (among other uses) refers to an action or a condition as coincidental with another action. Since the concordance of tenses is obligatory in Catalan, the form *va trer* also must have a past tense value. That is, the verbal tense in the subsequent clause shows that this occurrence is a fully grammaticalized form (cf. the discussion of occurrences of the GO-periphrasis with the verb 'go' in present tense in Section 2.5.3).

These considerations lead us to occurrences where the temporal relations between adjoining clauses and its verbal tense help us to infer contents that are not explicitly expressed. This is possible because, due to the obligatory concordance of tenses in Catalan, the tense of a verb in the subordinate clause changes in accordance with the tense of the verb in the main clause. Consider the following examples.

(92)

E puis **van**   **eléger**   *dos de cavall, e un adalill e un mogatèn, per qui*
       Aux.3Pl.Pres elect.Inf

<u>*es regissen*</u> *estrò haguessen cap.*

'And then they **elected** (= **go to elect**) two horsemen and an adalil, and a captain of the almugavars <u>to govern</u> them until they had a chief.'

(Munt II 121,44–45 – 122,1)

(93)

e aquí ell **va**   *fer*   *lo duc cavaller. E <u>con l'hac fet cavaller,</u>*
       Aux.3Sg.Pres make.Inf

lo duc dix davant tuit:

'and there he **dubbed** (= **?goes to dub**) the Duke knight. And <u>when he had dubbed him knight</u>, the Duke said before all:'

(Munt II 128,46 – 129,1)

(94)

E io llavores li **vaig**   **dir**   *que, ab l' ajuda de Déu, io hi* <u>*intraria*</u> *per*
                  Aux.1Sg.Pres   tell.Inf

purgar mos pecats;

'And then I **told** (= **?go to tell**) him that with God's help I <u>would enter</u> there to purge my sins.'

(Per 34)

In (92) the relative clause which follows the clause containing an occurrence of *anar* + Inf (literally: 'by whom they were governed') the form *es regissen* is conjugated in the imperfect subjunctive, which indicates that the main verb has a past tense value. In (93) the form *l'hac fet cavaller* 'he had dubbed him knight' refers back to the event expressed by the verbal form *va fer cavaller* in the preceding clause. The form *l'hac fet cavaller* is conjugated in the past perfect tense, which expresses an action, state, or event that was already completed before the start of another past action. That is, the event of dubbing the duke a knight was actually performed.[81] Finally, the text segment in (94) describes a fragment from a dialogue: as is clear from the broader context, the participants are at the same place, namely, at the gate of Paradise, that is, they are not likely to go anywhere. The past tense value of the form *vaig dir* is also revealed by grammatical properties of the verbal form in the subsequent clause: the content of the dialogue is expressed in indirect speech where the conditional form *intraria* '(I) would enter' is used. The conditional is used in Catalan to express futurity in the past. If the form *vaig dir* had a present tense value, in the subsequent clause futurity would be expressed with a simple future tense.

The use of verbal mood can also provide useful pieces of information. In the Catalan Example (95), the indicative mood[82] of the form *passà* 'passed' in the subordinate clause instead of the subjunctive mood, which would be used if the form *anà* preserved its original 'motion' meaning, signals that the verb *anar* does not retain its lexical meaning 'go' in that context.

(95)

mès-se la lansa denant, e **anà**   **ferir**   *lo cavaler alamayn de tal*
                    go.3Sg.PerfP   attack.Inf

vertut que la lança li <u>*passà*</u> *de l'altra part per mig lo cors*

'[he] held the lance out and **attacked** / (= **?went to attack**) the German knight in such a way that the lance <u>passed</u> right through him in the middle of his body'

(Desc II 55,25–27)

In Example (95), if the periphrasis referred to the motion with the intention of attacking, the following clause would have the verb in the subjunctive mood. The indicative mood suggests that the action described in the subsequent clause was not the aim but the consequence of the action in the GO-periphrasis. And it is more likely the consequence of the act of attacking than that of going. In other words, he did actually attack the German knight and the lance did actually pass through him. This grammatical feature reveals that the verb *anar* does not retain its lexical meaning 'go' and serves as evidence of a previous semantic change.

Naturally, the analysis of an even larger discourse segment may also help us to assign meanings. By broadening the context, we will also be able to make suppositions about how participants move in the space during the acts described in the text. To make such suppositions, often it is not enough to take into consideration the discourse context; we also need to extend the context with encyclopedic information. The following dialogue (96) (presented above in Section 2.6.3.2 as (38) without glosses) contains various occurrences of *anar* + Inf.

(96)

E quan ells haguen ab mi parlat així llongament, me menaren en una gran muntanya, e digueren-me que regardàs en aut e en avant, vers lo cel, e io hi

vaig	*regardar, e*	*van-me*	*demanar* de quina
Aux.1Sg.Pres	look.Inf	Aux.3Pl.Pres-Clit.1Sg	ask.Inf

color era ni cum me semblava ni ont era io. E io lus **vaig**   **responder**

Aux.1Sg.Pres   answer.Inf

que a mi me semblava color d'or e d'argent de la fornal ixit. E llavores ells

me	*van*	*dir:*
Clit.1Sg	Aux.3Pl.Pres	say.Inf

'And when they had spoken thus long with me, they led me to a great mountain and told me to look up and forward towards the sky. And I **looked** there and they **asked** me what colour it was and what it seemed to me and where I was. And I **answered** them that to me it seemed the colour of gold and silver come out of the furnace. And then they **said to me**:'

(Per 49)

Assigning a motion meaning to occurrences of *anar* in the above dialogue and thus assuming that we are dealing with a historical present usage would result in the following incoherent discourse, in which the participants are moving all over the place during the interaction:

??'And when they had spoken thus long with me, they led me to a great mountain and told me to look up and forward towards the sky. And I **?go (?to the sky) to look** there and they **?go (?to me) to ask** me what colour it was and what it seemed to me and where I was. And I **?go (?to them) to answer** them that to me it seemed the colour of gold and silver come out of the furnace. And then they **?go (?to me) to say to me**:'

To summarize Section 3.5, the context may contain several linguistic clues which guide us in assigning meaning to utterances in historical documents. The majority of them are based on the meaning of certain elements, either in the narrow or in the broad context (infinitives entering the periphrasis, nouns, adverbs, preceding or subsequent clauses, etc.). However, there are also certain grammatical characteristics which can help us, including verbal tense or verbal mood. The narrow context is often not enough to assign meaning, and the broad context should also be taken into account, including encyclopedic information (cognitive context), as well.

3.6 Argumentation in historical pragmatics: deciding between rival hypotheses

3.6.1 Introduction: metalinguistic considerations in historical pragmatic research

The aim of the present section is to demonstrate the usefulness of taking into account metalinguistic considerations in historical linguistic research. I repeat the related research question here for convenience:

    iv. What would be the proper notion of data in historical research?
    v. What argumentation techniques are used in historical pragmatics?
    vi. How can methodological considerations help us to decide between rival hypotheses during research?

Metalinguistic reflection refers to an examination of the process of linguistic research from outside and a continuous monitoring of the correctness of our methodology. We examine what pieces of data we are relying on, and we also analyse the process of argumentation. Throughout the presentation of my research in Chapter 2, I referred to relevant linguistic units taken from the corpus using the terms *occurrence* or *example* and I avoided the term *data*, although in linguistics literature all three terms are traditionally used as synonyms. Methodological considerations shed light on the significant difference between them. The metalinguistic analysis of the argumentation process also reveals problems originating from an intuitive approach. Linguists usually apply the methods and argumentation techniques accepted in the linguistics area they work in automatically, without conscious reflection, and sometimes they do not make explicit every assumption that plays a role in their argumentation. Statements that are not explicitly expressed in the argumentation are so-called latent background assumptions (Kertész and Rákosi 2012: 89–91). They are nevertheless needed to secure the plausibility of conclusions in certain cases. Metalinguistic considerations can also help us to decide between competing hypotheses. When examining data and methods in historical pragmatics and analysing the research process from a methodological point of view, I rely on a model of scientific theorizing, the so-called p-model as presented in Kertész and Rákosi (2012; 2014b). In the following section we will begin by examining theoretical considerations on the notion of data.

3.6.2 A model for scientific theorizing: the p-model

3.6.2.1 The notion of data

The term *historical data* usually appears in the relevant literature as a synonym for occurrences found in a collection of texts compiled for historical research. The central role occurrences of a certain linguistic unit found in historical documents play in historical research is unquestionable. It is, however, important to establish whether these occurrences really are data in themselves, and also to identify the actual role they play in the research process. My considerations about data concern mainly corpus data, since historical research is based predominantly on data gained from historical documents. Written corpora have always played a pivotal role in historical linguistics. Therefore, in the present study I concentrate on corpus data. Fischer (2004: 730) maintains that corpus data have the central role in historical linguistics, and historical documents are the only firm source of knowledge for historical linguists (see above in Section 3.2). Lehmann (2004: 192, 196ff., 201) also states that data taken from corpora are the most reliable type of data. According to Lehmann, corpus data, due to their 'discovered' nature, enable more objective research and a more restricted possibility of the manipulation of data, since they are independent of the researcher. A discovered corpus is varied, offers surprises, and makes unexpected findings possible. Fischer's and Lehmann's optimistic approaches are also manifested in historical research practice: studies relying on a large corpus are accorded high prestige. However, collecting and using corpus data raises several methodological questions, and we should ask whether a corpus can actually be considered a perfectly reliable data source which leads to more objective results (cf. Kertész and Rákosi 2012: 173–175). In order to answer this question, I rely on the p-model as developed in Kertész and Rákosi (2012, 2014b).

The p-model emphasizes the usefulness of taking into account metalinguistic considerations in order to decide between competing hypotheses in linguistic research and to judge the reliability of certain claims. The name *p-model* as used by Kertész and Rákosi refers to the central idea of this approach: the authors consider linguistic theorizing to be a process of plausible argumentation that leads from plausible data to plausible results (Kertész and Rákosi 2012: 54). According to this approach, the authors elaborate a novel notion of data, which reflects the idea that pieces of information at one's disposal during research are not certainly true, but more or less insecure, i.e. merely plausible. The authors do not use the term *data* in the traditional sense, as discussed above. The p-model defines data as plausible statements originating from some direct source. Data have a double structure: they combine an information content and a plausibility value, which is assigned on the basis of some direct source. Quoting Rescher (1976: 6), Kertész and Rákosi (2012: 64) mention various direct sources on the basis of which we can assign a plausibility value to data. The main types are the following: individuals' accounts, depersonalized historical sources, one's own observations, intellectual resources, and methodological principles. The initial plausibility value of a piece of data is assigned on the basis of the reliability of its source, but it may be re-evaluated during the argumentation process, if the reliability of the source is modified. The p-model also enables a statement to be plausible on the basis of a source and implausible on the basis of another at the same time.

Let us provide an example. Concerning the current use of Catalan *anar* + Inf (presented in Section 2.1.2.2), Pérez Saldanya (1996: 101–102) claims that the construction cannot be used with the verb *deure* 'must', but Juge (2002) provides evidence to the contrary. The difference in their opinion concerns the following statement (S), assumed to be true by Pérez Saldanya (1996), but rejected in Juge (2002): '(S) The combination of the construction *anar* + Inf with the infinitive *deure* 'must' is not grammatical in current Catalan.'

Inasmuch as this problem is relevant in research, the statement (S) can function as a piece of data, whose initial positive plausibility value stems from Pérez Saldanya's (1996: 101–102) work as a direct source. However, if we also take into account Juge (2002) as a further source, we find an occurrence of *anar* + Inf with *deure*. On the basis of this occurrence provided in Juge (2002), a piece of data can be retrieved which has a double structure: it combines the information content (according to which the combination of *anar* + Inf with the infinitive *deure* can be found in Juge's corpus and is grammatical in current Catalan) and a plausibility value (in this case, it is plausible on the basis of Juge (2002) as a direct source). That is, (S) turns out to be implausible on the basis of Juge (2002) as a direct source. As I have mentioned above, the p-model also enables a statement to be plausible on the basis of one source and implausible on the basis of another at the same time, as frequently occurs during scientific research. We can solve this problem in different ways. We can revise or call into question the reliability of Pérez Saldanya (1996) as a direct source concerning (S). We can argue that Pérez

Saldanya has not taken into consideration enough examples of the construction *anar* + Inf, and Juge (2002) relies on a wider range of occurrences, since we can deem it a more reliable source in this respect. Nevertheless, it can also happen that new information originating from further sources undermines the acceptability of Juge (2002) as a data source, revealing that the occurrence provided in it is a performance error, a slip of the tongue, etc.

Taking into account the above considerations, the three types of data distinguished in Jucker (2009: 1611, 1615–1619; see 3.2) – data relying on the language users' intuition, data based on everyday language use or compiled with the philological or corpus method, and elicited data – can all be treated in the same way in the p-model. They can all be regarded as statements about occurrences studied by the researcher, which, beyond their informational content, also have a plausibility value. Their initial plausibility value is assigned on the basis of some direct source: linguistic intuition, a written or oral corpus, and some elicitation technique, respectively.

3.6.2.2 Corpus data in historical research

In the investigation of former language states, both historical linguists and historical pragmaticians have to restrict themselves to written sources and indirect evidence.[83] Working with corpora, however, has both disadvantages and advantages. In Kertész and Rákosi's approach (presented in the previous Section 3.6.2.1), a piece of data based on a corpus is a statement with a plausibility value which states the presence of a linguistic unit in a corpus or captures some of its characteristics. Kertész and Rákosi (2012: 173) show that it can have various structures: e.g. 'The utterance U containing the linguistic phenomenon P can be found in corpus C', or 'The structure S can be identified in corpus C and it had the characteristics X'. Corpus data can be interpreted as plausible statements originating from a direct source; that is, they cannot be considered to be certainly true statements. The authors (Kertész and Rákosi 2012: 173; cf. Forgács 1993–1994) emphasize that the reliability of corpus data is influenced by several factors, among others the reliability of the historical document, the researcher's intuition, the difficulties in the identification and interpretation of the linguistic phenomenon under study, etc. Kertész and Rákosi (2012: 173–175) discuss four problems concerning a corpus as a data source that I present here, adding my own observations and adapting them to historical research. They are the following. (1) One has to decide whether a sentence from the corpus is part of the language one is studying (i.e. whether it is grammatical or not). (2) The linguist has to select data; corpus data are therefore inevitably theory-dependent. (3) Using corpora requires the active involvement of linguistic intuition. And, finally, (4) the question of positive and negative evidence arises in connection with corpora.

As a consequence of these problems, information based on the use of a corpus cannot be regarded as certainly true when involved in the process of scientific argumentation. The traditional approach to historical data does not reflect these uncertainties. Thus, hereafter I use the term *data* according to the notion provided in the p-model, because this novel concept of data helps us to analyse the problems which arise in connection with the use of corpora. Let us consider the four problems mentioned above in more detail.

1. During the use of a corpus, as Kertész and Rákosi (2012: 174) emphasize, researchers have to decide whether a piece of information gained from it is part of the language one is studying or not. In other words, we have to make a distinction between grammatical forms and, for instance, misspellings. This difficulty manifests itself more seriously in historical research. On the one hand, while in synchronic research there is a possibility of testing with the help of native language speakers whether some 'suspicious' occurrence found in the corpus can be considered part of the language under study or not, historical linguists lack this option. On the other hand, in historical research it is not enough to decide whether an occurrence belongs to the language under study or not; an assessment of which language stage it belongs to is also necessary. Forms belonging to earlier and later language stages can appear in the same historical text – in other words, a historical text does not necessarily reflect (only) the language state of the period when it was written. In addition to grammatical forms which were in use in the given language state and accidental misspellings, historical texts can contain (a) archaic forms, (b) words, corrections, remarks, or commentaries inserted by a later copyist or translator of the text, and (c) words, strings of words, or even whole paragraphs written in another language. Let us illustrate each case with an example taken from the present research. (a) Bruguera (1981), on the basis of a thorough linguistic analysis, characterizes the Chronicle of James I as a linguistically archaic text, although it was composed not long before Desclot's Chronicle. The tense of the occurrences of *anar* + Inf reflects this archaic character. (b) As presented above (see Section 2.5.2), Desclot's Chronicle contains two occurrences of the *anar* + Inf construction with the preposition *a* 'to'. Both occurrences appear in a title, and we know that titles were often added later to historical texts, and, as a consequence, these latter occurrences should be considered as belonging to a later language stage (see Pérez Saldanya 1996: 74). (c) In chapter 272 of Muntaner's Chronicle there is a poem of epic character in Provençal.

2. The following difficulty mentioned in Kertész and Rákosi (2012: 174) is that we should decide which strings of words in the corpus are relevant to our research topic; in other words, we have to select potential data; we have to identify the

phenomena we are looking for in our corpus. These decisions are determined by the theoretical framework we are working in: the analysis of the text according to a theory and the identification of relevant structures lead to the theory-dependence of data (cf. Lehmann 2004: 183; Fischer 2007: 17, among others). As Kertész and Rákosi (2012: 171) emphasize, grammaticality judgments presuppose some theoretical framework, which is used as a point of reference when judging the grammaticality of a sentence. In a similar way, a previous grammatical analysis is needed to decide whether a form found in the corpus can be considered an instance of the structure we are studying. For instance, I have not included in my study occurrences of *anar-se'n* 'to leave, to go out' used in a purposive construction, because I judged it to be a different construction and irrelevant in the research on the grammaticalization of *anar* + Inf, since in these occurrences the motion verb always conveys its literal meaning. In contrast, Bruguera (1981: 33–35) mentions among his examples occurrences of this construction too (see his examples 6, 16, 19, 38, 39, 46, and 50).

3. When processing a corpus, the researcher unavoidably relies on his/her linguistic intuition (Kertész and Rákosi 2012: 173–174). The role of a researcher's linguistic intuition while processing corpora can be observed in Bruguera's (1981: 38) method of revealing the meaning conveyed by occurrences of the Catalan *anar* + Inf construction, compiled from the Chronicle of James I. In the case of each particular occurrence of *anar* + Inf, Bruguera (1981: 38) decides whether the verb *anar* 'go' still conveys its original, lexical meaning or can be interpreted as a grammaticalized form relying on the following thought experiment. He constructs minimal pairs, that is, he translates each occurrence to Modern Catalan in two different ways. In the first interpretation, he translates the verb *anar* as a main verb, with the 'motion' meaning; in this case in current Catalan the preposition *a* is obligatory (cf. Section 2.1.2.2). In the second interpretation, he translates the occurrence with the periphrastic perfective past of Modern Catalan, that is, he regards it as an example of the periphrastic past tense. Bruguera maintains that the observation of the participants' situation in the space and the act described in the infinitive in relevant contexts suggest a motion which leads to a different scene. However, the text resulting from the second translation does not express this change of scene. Taking into account this information, Bruguera argues that if we accept the second interpretation, the meaning of the translated version does not coincide with the original to the same degree as if we accept the first translation. He concludes that the correct interpretation of the occurrences from the Chronicle of James I is the literal interpretation, in other words, the grammaticalization of the construction has not yet started in the language state reflected in this text. The methodological principle I have followed when guessing meanings,

according to which we have to assign the source meaning as long as is possible, also presupposes that the researcher is able to decide whether the literal meaning is the correct interpretation in a given context.

In the course of historical research, and also in the case of Bruguera's (1981) method, the linguist has to judge phenomena of a language state in which he/she does not have proper competence (cf. Dér 2004: 191–192). Forgács (1993–1994) discusses the possibilities and limits of developing what has been called substitute competence for this purpose. Discussing the notion of historical data, Dömötör (2012: 45) refers to a similar concept, but she mentions the terms *acquired intuition* and *historical language sense*. Using such a language sense, the problems originating from the use of the researcher's linguistic intuition turn out to be even more serious.

Forgács (1993–1994: 18) refers to historical corpora as closed corpora, following Greule (1982: 72). A closed corpus (in contrast to an open one) is characterized by the absence of informants who speak the language of the corpus natively, and where the researcher does not have proper competence in the language of the corpus. As a consequence, it is not possible to clarify the uncertainties related to various pieces of information obtained from the corpus. In order to make research based on a closed corpus more reliable, the linguist should evolve a substitute competence, and in this way he/she can try to fill the gap mentioned above. It is easy to see that such substitute competence cannot rival native language competence.

The notions *sense of system* and *sense of the grammatical system* provided in Fischer (2007: 15) can be connected to *substitute competence*. Fischer (2007: 15) holds that we should investigate not only the proper topic of our research, but 'other forms and structures that are not strictly cognate', too. In order to do this, a certain sense of the grammatical system of the period is desirable in historical research, an ability which corresponds to the notion of substitute competence.

The phenomenon of using a substitute competence can also be related to the issue of indirectness. The fact that the person interpreting the text does not possess native language competence in the language state reflected in it means that the access to the language state in question is more indirect. On the one hand, the reliability of the substitute competence is questionable, and on the other hand, the interpreter's competence of her/his native language may influence the interpretation. Is it possible to eliminate the interference between native language competence and substitute competence, and if so, to what degree? Fischer (2007: 18) essentially refers to this interference when she draws attention to the possibility of interpreting historical texts from the point of view of the modern language system. This is more likely to happen when the linguistic form has not suffered considerable changes on the surface. The risk of interference of competences can be decreased by developing a historical grammar sense: in other words, broadening our knowledge about the grammatical system of the language state we are studying. However, it seems improbable that the problem can be totally eliminated.

A similar interference of competences can be observed when competences related to the written and the oral language influence each other. Fischer (2007: 41) claims that if we compare written constructions in the modern language and earlier periods, then 'we may be comparing apples with pears', since the written language of earlier periods is often closer to an oral style. Later, when the written mode fully develops and becomes conventionalized, it acquires special characteristics which contrast with the oral style. However, grammaticality judgments often take as a starting point the written language, and we judge as unacceptable structures and phenomena which are usual in the oral style. Thus, following Biber and Finegan (1994), Fischer (2007: 13) draws attention to the fact that if we wish to compare a homogeneous group of texts, we should not rely on genre but rather on linguistic indicators (such as, for instance, the absence/presence of passive constructions, nominalizations, subordination, first or second person pronouns, etc.).

4. Finally, the last question which arises in connection with the use of corpora is the problem of positive and negative evidence. According to Kertész and Rákosi (2012: 178), evidence is a datum 'whose function is to *contribute to the judgment and comparison of the plausibility of rival hypotheses*' (emphasis in the original).[84]

In the case of positive evidence, an instance or instances of a structure studied by the linguist appear in the corpus. The presence of occurrences, in a first approximation, makes plausible the statement that the linguistic construction in question can be considered grammatical in the language state represented by the corpus. Kertész and Rákosi (2012: 19–20, 69, 173–175) draw attention to the problem that this is not necessarily the case, since the corpus may contain typos, slips of the tongue, unreliable, dubious information, faulty forms (cf. problem 1). It is worth adding that even if a linguistic form is present in a corpus, whether it represents the phenomenon under study depends on the subjective judgment of the researcher (cf. problem 2).

Since a (historical) text can combine parts which reflect different language states, the interpretation of positive evidence is even more problematic. In order to be able to assign a plausibility value to a statement about the acceptability of a linguistic structure present in a corpus, we have to know to which language state the part of the text in question belongs. The reliability of our judgment depends on whether we judge correctly which language stage the text represents. It is essential that the researcher should be able to distinguish between parts of the text belonging to the language stage he/she is studying and parts which belong to another language stage or can be considered misspellings, incorrect forms, later remarks, forms belonging to another language, etc. This is essential, since data construed on the basis of occurrences in these latter parts of the text cannot be considered positive evidence regarding the phenomenon under study.

The problem of how to interpret negative evidence is even more serious. In the case of negative evidence, the corpus does not contain any occurrence of a certain form or construction, and, in a first approximation, its absence makes the statement that the construction is ungrammatical or faulty in that language (language state) plausible (see Kertész and Rákosi 2012: 19–20, 69, 174). The problem of negative evidence also arises in the research presented in Chapter 2. For instance, in the case of constructions combinable with the prepositions *a* or *de*, some infinitival combinations were attested with one of the prepositions, but not with the other one, or without it. In this case it is quite reasonable to suppose that examples of the missing combination would occur in a larger corpus (cf. also considerations in Chapter 2 n. 40). It can also be the case that they are present in our corpus too, but we have overlooked them. Therefore, the absence of a certain construction in our corpus cannot automatically be regarded as negative evidence; in other words, it does not make implausible the statement that the linguistic construction in question can be considered grammatical in the language state represented by the corpus.

Let us illustrate these problems with the investigation presented in Bybee (2003/2005: 605–614; discussed above in Section 2.6.3.1 and repeated here for convenience). This concerns the process by which the English verb *can* has become an auxiliary. Bybee applies the type frequency counting method on a corpus of Old and Middle English texts, considering 300 occurrences of the construction with *can*. She makes a distinction between verbs of intellectual states or activities, verbs of communication, and verbs of skills, since these are the verb types which typically appear with *can*. The results from the Middle English corpus concerning different semantic groups of infinitives which can appear together with *can* are the following:

1.   Verbs of intellectual states or activities: 52 tokens, 18 types

     high token frequency: *see* 12, *deem* 6, *understand* 6, *espy* 'discover' 5

     type frequency: 18 distinct verbs

2.   Verbs of communication: 102 tokens, 31 types

     high token frequency: *tell* 30, *say* 29, *devyce* 'describe' 8

     type frequency: 31 distinct verbs

3.   Verbs of skills ('know how to'): 26 tokens, 18 types (Bybee 2003/2005: 611–612).

Bybee compares these Middle English proportions with proportions from Old English. The comparison shows that, although infinitives belonging to these groups could also be used with *can* in Old English, in Middle English more infinitives are documented in each group, and furthermore, the token frequency of some combinations in the first two groups of infinitive types increases. At the same time, new infinitives appear with *can*.

According to Bybee, both token frequency and type frequency are relevant in grammaticalization research, since both contribute to semantic bleaching. Bybee considers it remarkable that she found, for instance, occurrences of 31 different lexemes within the group of verbs of communication in the Middle English corpus, in contrast to the more limited presence of distinct verbs belonging to the same type in the Old English corpus. The absence of verbs that are present in the Middle English corpus, but do not appear in Old English texts, functions as negative evidence when describing the use of the construction in the Old English language state. Their absence suggests that these combinations were not yet in use in that period. In contrast, it seems probable that she would not assign importance to the absence of the combination of *can* with another verb which has a cognate meaning with one of the verbs occurring in the Middle English corpus. For instance, the finding that both the type and token frequency of verbs of communication are very high in the Middle English corpus makes it plausible that the absence of a combination of *can* with a further verb of communication is only accidental, and with the broadening of the corpus we could find occurrences of this phrase, too.

Similarly, concerning the research presented in Chapter 2, we consider it significant if we do not find any occurrence of *anar* + Inf with the preposition *a* 'to' in a certain corpus. Since the spread of prepositional forms correlates with stages of grammaticalization, we interpret the absence of prepositional forms as evidence for the assertion that the grammaticalization of the construction has not yet started in that language state. However, it is possible that in a broader corpus we could find a text belonging to the language state under study which would contain prepositional forms, and in that case we would modify the dating of the grammaticalization of the construction. The lack of occurrences of the construction with the preposition *a* can be regarded in some cases as negative evidence. In contrast, the absence of prepositional occurrences of combinations of the verb *anar* and a certain infinitive in a corpus which represents a language state in which the prepositional construction has already evolved as a type would not be regarded as negative evidence (cf. Chapter 2 n. 40). Since attestations of prepositional occurrences with other infinitives suggest the existence of the prepositional construction as a type, we suppose that the absent combination may turn up in a larger corpus. To summarize, we consider the absence of prepositional combinations as negative evidence in some cases, but not in others. Our decision making is also influenced by our knowledge of the language system we are studying.

Naturally, the more extended our corpus is, the heavier the negative evidence. However, the broadening of the corpus has its limits in historical research (and elsewhere); therefore, hypotheses based on negative evidence can never be regarded as necessarily true.

The absence of a construction or phenomenon in a text can occasionally be interpreted as a characteristic of the genre. Colon (1959/1978) examines about a hundred texts, and only in every second text does he find occurrences of the Catalan *anar* + Inf construction. Colon (1959/1978: 120) suggests that this quantity of text is enough to observe certain correspondences concerning the genres in which the construction was used in a certain language state.

Accordingly, it is impossible to decide with absolute certainty whether the absence of a construction is something incidental or can be attributed to the grammatical characteristics of a certain language state, the length or style of the text, or the author's individual style. As an example of this, see the description provided in Soldevila (1963–1968) about the use of the Catalan *anar* + Inf in written language. Soldevila observes that in Muntaner's medieval Chronicle, more and more occurrences of the construction *anar* + Inf appear towards the end of the text, and he explains them – constructing a parallel with his own language use, since he uses the periphrastic perfective past for a stylistic reason, in order to lend variety to his style – that the author became enthusiastic about the use of the construction; in other words, he attributes these occurrences which accumulated in the same context to Muntaner's individual style.

Considering all these findings, it seems reasonable not to regard any corpus, and, naturally, any historical corpus, as a perfectly reliable source of information. Corpus data cannot be regarded as certainly true but only as plausible statements, and Kertész and Rákosi (2012: 173–175) emphasize that corpora should be supplemented with further sources when explaining linguistic phenomena (see also Kertész and Rákosi 2008). To sum up, neither the absence, nor the presence of a certain form in a corpus helps us to decide with certainty whether the construction in question was in use in the language state under study.

3.6.2.3 Historical corpus data as data from an integrated data source

So far I have discussed corpus data as if they were data based exclusively on a corpus. Using the third problem concerning corpus data as a starting point, we can make further observations. Data based on a corpus are always based on the intuition of the linguist as well, since the interpreter's intuition necessarily plays a role in processing corpora. The first case of the use of linguistic intuition is grammaticality judgments. When processing a historical text, the researcher first has to decide whether a form found in a corpus is a grammatical or a faulty form. This is a relatively simple case of using linguistic intuition, but there may even be differences between different researchers' judgments. Moreover, if the meaning of a certain form or difficult to access levels of meaning such as those investigated in pragmatics research are involved, variance becomes even more frequent and serious.

Bruguera's (1981) method (discussed above in Section 3.6.2.2, and repeated here for convenience) is a good illustration of the role the linguist's intuition plays when interpreting occurrences found in a historical text. Bruguera (1981: 38), when revealing the meaning conveyed by occurrences of the Catalan *anar* + Inf compiled from the Chronicle of James I, relies on the following thought experiment. He translates each occurrence of *anar* + Inf to Modern Catalan in two different ways. In the first interpretation he translates the verb *anar* as a main verb, with the meaning 'go'; in this case the preposition *a* is obligatory between the verbal elements (cf. Section 2.1.2.2). In the second interpretation, he translates the occurrence with the periphrastic perfective past of Modern Catalan, that is, he regards it as an example of the periphrastic past tense. Bruguera argues that the translated version according to the second interpretation does not express the change of scene present in the original context. This method is very useful when examining meaning nuances in historical texts, and, as can be seen in the analyses of examples, I have also applied it. As regards the change of scene, there are obvious cases,[85] but in other cases the motion in space is so limited that it is not clear whether the change of scene should be included in the interpretation or not (see Section 2.6.8, point 2). The study of the possible presence of inferential contents in the context of written discourses is even more difficult. This topic has been addressed in Section 3.4.2.

To sum up, the source of historical data conceived of as a plausible statement in a first approximation is a corpus. However, since, on the one hand, the direct source of a piece of historical data is not only the historical document itself, but also the researcher's intuition, and, on the other hand, the plausibility value of corpus data can be influenced by the plausibility value of data from other sources, historical corpus data should be conceived of as data from an integrated data source. Consequently, instead of the term *historical corpus data* it would be more felicitous to use the term *partially corpus based historical data* or *data construed on the basis of a historical corpus*.

Finally, using various examples, let us investigate how this integrated source is created. I discuss examples taken from the research process presented in Chapter 2 in order to illustrate what pieces of information are merged together in the plausible statements traditionally called corpus data – in our case, historical corpus data.

Example (a)

Historical documents pose a number of challenges for those seeking a modern interpretation. Some difficulties in the interpretation originate from the writing traditions of particular eras. In historical documents from before the standardization of written language, punctuation is frequently lacking, and documents may contain indecipherable handwriting, nonstandard spelling, obscure abbreviations, inconsistent punctuation, etc. Linguistic forms may appear in various spellings, even within the same text. For instance, in the Spanish texts of the research presented in this book, the verb meaning 'receive' appears with the following different spellings: *recebir*, *reçebir*, or *rresçebir*. In order to establish that all these spellings represent the same verb, the researcher should use his/her linguistic intuition. Moreover, while in some cases different spellings are due to the absence of a written standard, in other cases they can be considered genuine misspellings, or can reflect some ongoing sound change. How can we decide which is the most likely case? As is already evident in this very simple example, we should take into account the following factors in order to be able to make a decision: the researcher's linguistic intuition, information concerning further occurrences, the written form of the sounds affected in other words, the place of the 'error' or spelling difference within the word, the phonetic/phonological system in the language state under study, the phonetic/phonological system in earlier or later language states, common pathways of phonetic/phonological changes described in relevant literature etc. Historical phonologists may construe a usable datum on the basis of these occurrences if they consider all these pieces of information.

Example (b)

As for the interpretation of ambiguous forms, let us consider further examples. Difficulties may originate from some historical documents which do not use punctuation. For instance, in medieval Catalan manuscripts which do not mark word stress there is a formal coincidence between the present and preterit tense of the 3Sg forms of first conjugation verbs. For instance, the forms *parla* '(he) speaks/is speaking' and the end-stressed *parlà* '(he) spoke' can both appear as *parla* in medieval texts (see Bruguera 1981: 31). So, it is difficult to decide whether it should be interpreted as a present or a preterit verb. Similarly, there was a formal coincidence between the present and past forms in the first person plural of verbs belonging to all the three conjugations. Because of this, it is hard to decide whether such a form should be interpreted as a present or past tense verb (Bruguera 1981: 31). For instance, the 1Pl form of the Catalan verb *anar* 'go' (*anam*) can be interpreted as either a present or a perfective past form ('[we] go/are going/went') in the medieval language state (see Juge 2006: 320). However, its interpretation in certain places in a historical text plays a fundamental role in forming hypotheses about the history of the Catalan *anar* + Inf construction. Instances of the form in question from the Chronicle of James I are interpreted as historical present verbs in Coromines (1980–1991), while as perfective past verbs in Bruguera (1981). Bruguera argues that beyond these morphologically ambiguous forms there are no other verbs conjugated in the present in relevant contexts of the text. In similar cases, when constructing a piece of historical data, we should take into account results of earlier historical research on the same topic, the morphology of neighbouring verbal forms, the broad context of the occurrence, information concerning the writing practices of the period, and information and theoretical considerations concerning discourse organization, among others.

Example (c)

If we are interested in the distribution of the constructions *anar* 'go' + Inf and the prepositional *anar* 'go' *a* + Inf, the following pieces of information, among others, can be taken into consideration in order to judge the status of occurrences with the preposition *a*: the place of the occurrence in the text, information based on the comparison of different manuscripts or translations of the same text, encyclopedic information about historical documents.

On the basis of the above considerations, we can summarize that historical corpus data should be conceived of as data from an integrated data source. The historical corpus as a data source should be complemented with further sources, including the following: the researcher's linguistic intuition, results of earlier investigations on the same topic and other relevant issues, and the theoretical framework itself. Consequently, instead of *corpus data*, the appropriate terms to apply would be *partially corpus based data* or *data construed on the basis of a corpus*.

When construing a piece of historical corpus data, we rely on other pieces of information as well, by taking into account the following: various occurrences of the linguistic form under study, the place of the occurrence, the broad context, further relevant contexts, the whole language stage under study, earlier and later language stages, encyclopedic information concerning the historical documents in our corpus, writing traditions of the periods, further manuscripts, translations, data concerning other languages, typological data, etc.

After considering theoretical questions concerning the notion of data, let us continue with the argumentation process in linguistic research.

3.6.2.4 The argumentation process in linguistics according to the p-model

According to the p-model, the initial p-context of argumentation involves, among others, the existing accounts of the phenomenon under study and assumptions originating from our own data sources (e.g. historical corpora, consultation with native speaker informants, inferences, etc.). However, as Kertész and Rákosi (2012: 122) claim, the p-context is not simply the amount of previously raised hypotheses concerning the object of research, but is more complex, because it covers the following:

(a) the accepted methodological norms, for instance, the criteria of the acceptability of the statements, the permissible type(s) of inferences, methodological rules pertaining to the relevant characteristics of the accepted inference types (the useable logical, semantic etc. systems, the methods of the translation of the sentences from natural language to some logical and semantic system), the methods of the treatment of contradictions, the criteria for the judgement of the reliability of sources etc.;

(b) the direct and indirect sources (inferences) judged as reliable on the basis of the criteria in (a);

(c) statements and their relevant characteristics at one's disposal according to (a) and (b), or some subset of this set.

It is possible that the starting p-context contains statements which are plausible on the basis of some source, but implausible on the basis of others, and during research we have to decide between these rival hypotheses. Kertész and Rákosi call this phenomenon *p-inconsistency*, which is a kind of *informational overdetermination*. According to Kertész and Rákosi (2012: 130), 'the p-context is called p-inconsistent if there is a statement p such that both p and $\sim p$ have a positive plausibility value according to any source in C'. The p-context in other cases is incomplete, or informationally underdetermined, if there are statements which cannot be regarded as either plausible or implausible, since the sources do not provide enough information to assign a plausibility value to them. If the statement is considered to be relevant in the research, we can attempt to eliminate the p-incompleteness by involving further sources. A p-context can be informationally over- and underdetermined at the same time. In order to resolve a p-problem, the transformation of the initial context is needed. Kertész and Rákosi (2012: 134) regard linguistic research as a process of plausible argumentation which is 'the gradual transformation of a p-problematic p-context into one that is no longer (or at least less) p-problematic'.

The first step in the development of a solution to a scientific problem is the extension of the starting context of research. Then, the extended context has to be coordinated, that is, the plausibility values of statements stemming from the old and new sources, their relevant characteristics, and old and new pieces of information related to the reliability of the sources have to be compared. Finally, the coordinated context has to be modified, working out a re-evaluated version of the starting context (Kertész and Rákosi 2012: 136–143).

In the following, relying on the p-model as presented above, I examine hypotheses originating from previous literature from a metatheoretical point of view with the aim of deciding between them.

3.6.3 Argumentation in historical linguistics research practice

3.6.3.1 Introduction: traditions of argumentation in historical research

Since the initial context involves accepted methodological norms in a given research area, let us begin by considering the major factors affecting the acceptability and plausibility of hypotheses in historical linguistics. In historical linguistics we should assign a high plausibility value to a statement if:

1. sources traditionally accepted in historical linguistics (e.g. historical corpora, a properly reconstructed language system, etc.) are used in the research,
2. research methods and argumentation techniques traditionally accepted in historical linguistics (e.g. analogy-based argumentation, linguistic reconstruction methods, frequency counting, etc.) are used in the research,
3. the analysis is supported by, or at least does not contradict, accounts of other linguistic levels; in other words, a historical pragmatic account is more plausible if it accords with other morphological, syntactic, etc. descriptions,
4. it is supported by, or at least does not contradict, cross-linguistically recurrent historical linguistic tendencies detected in previous, reliable research.

I have already presented data sources of historical research (see Sections 3.2 and 3.3); let us now discuss argumentation methods, although I will not provide an exhaustive list. We will only consider two key notions in historical linguistics which have also played a primary role in the research presented in Chapter 2: these are frequency and analogy.

3.6.3.2 Frequency

Frequency is a central notion in historical linguistics in two respects: as a phenomenon and as a method. On the one hand, several historical phenomena are associated with the high frequency of the linguistic units concerned, and on the other, frequency counting is a central method for investigating various historical phenomena. Nevertheless, it must be noted that the term *frequency* is not generally used as a statistical notion, but intuitively. Bybee (2003/2005) specifically deals with the role of frequency in semantic change, but explicit definitions for *frequent* and *rare* are missing, even in her work. In her analysis of the process by which the English verb *can* has become an auxiliary, the wording is as follows: 'use with inanimate subjects is *extremely rare*: only 12 examples are found in the corpus of 300' (Bybee 2003/2005: 613, the emphasis is mine). This problem is partly due to the fact that in the research on those language stages from which we possess a limited number of written documents, a statistical analysis is not possible. Hence, broadening our data sources has a great significance in historical research.

The definition of grammaticalization as 'the process of automatization of frequently occurring sequences of linguistic elements' (Bybee 2007: 969) shows the importance of frequency in grammaticalization as well, especially in relation to the generalization of meaning. Some linguists consider the generalization of meaning as a precondition for more frequent use. Others think the opposite: it is frequent use which facilitates the generalization of meaning (Dér 2005: 48, cf. Ladányi 1998: 420–422), i.e. according to this concept of frequency, it has a cognitive impact. Actually, both claims seem to be maintainable. As Bybee (2003/2005: 602) points out, 'frequency is not just a result of grammaticization, it is also a primary contributor to the process.' She argues that one of the main characteristics of grammatical units is that, compared to lexical morphemes, they occur extremely frequently, and this frequency is not simply a result of grammaticalization, but is its main motivating power. Examining the effects of repetition on constructions,[86] Bybee (2005: 10–11) finds that, while low levels of repetition lead only to conventionalization, extremely high frequency leads to the grammaticalization of the new construction, and that 'the changes that take place in grammaticization are conditioned at least in part by high frequency of use' (Bybee 2005: 11). These changes are the following: phonological reduction, the gaining of autonomy of the new construction, loss of the original meaning, and addition of a new, inferential meaning from the context.

Let us consider the other concept of frequency. Frequency counting is a powerful device of historical linguistics methodology that provides quantitative data and is applied in order to investigate the above-mentioned cognitive processes. According to Bybee (2003/2005: 604), counting token frequency and type frequency (discussed above, under 2.6.3.1) are two relevant methods of counting frequency in linguistic studies. High type frequency corresponds to a higher degree of grammaticalization. Type frequency and token frequency are closely related, since the increase in the number of lexical units which can co-occur with a certain grammaticalizing linguistic item triggers an increase in token frequency, which leads to several formal and functional changes (Bybee 2003/2005: 605).

The question arises of how frequency which affects cognitive processes in grammaticalization can be detected with the methods used in linguistics. The notion of the frequency of a linguistic unit in a certain language stage is only a theoretical construct. Moreover, the research question we are interested in and the theoretical framework we work in determine what we count the frequency of. In this sense, quantitative data obtained from frequency counting are theory-dependent. After formulating the research question, a further difficulty arises: whether we are able to compile a representative corpus. The representativeness of the corpus also depends on theoretical considerations and is very difficult to achieve, if it is at all possible, even for a current language stage; consequently, it is even more difficult for those language stages for which we do not have enough written documents. As a consequence, it remains questionable whether the frequency of a construction or phenomenon in a corpus reflects its actual frequency in the given language stage. In the case of language stages for which it is impossible to compile a sufficiently large corpus, the interpretation of negative evidence is very problematic, because it is probable that a given linguistic structure is missing only due to the paucity of accessible texts. In spite of all these considerations, historical research practice relies on the optimistic idea that data originating from – in some cases rather limited – corpora is suitable for researching several historical phenomena. However, the uncertainties considered above should be taken into account when assigning plausibility value to data from these sources.

3.6.3.3 Analogy and reanalysis

Another widely used technique of argumentation in historical linguistics relies on analogy. Numerous historical changes discussed in the literature are attributed to analogical processes (Hock 1991: 167). In historical linguistics, the term 'analogy' traditionally covers processes whereby, on the basis of a pre-existing structural pattern in a language, corresponding new patterns are introduced, or old patterns are replaced (Jeffers and Lehiste 1979: 60; cf. also Bynon 1983). Analogical processes are at work at every linguistic level. For instance, as is widely maintained in historical linguistics, some changes in phonetic structure are influenced by factors which are not phonetic in nature. One of these is analogy, whose main function is to make the form of lexical units which are similar from a morphological, syntactic, or semantic point of view, phonetically and morphologically similar, as well. In morphology, when a grammatical distinction is lost in some morphological classes as a result of phonetic changes, the pattern of identity is usually extended by analogy (Jeffers and Lehiste 1979: 71). Hock (2003/2005: 455) presents a wide range of phenomena which have been considered analogical, namely, four-part analogy, levelling, morphophonemic extension, blending, contamination, recomposition, and folk etymology. He concludes that different analogical changes constitute points in a continuum of changes which may be considered analogical, and their differences are a consequence of the differences in the domain(s) in which they can apply.

Literature on grammaticalization has also recently addressed issues concerning the role of analogy in language change. Analogical change and grammaticalization are two different processes, but they can co-occur and interact in particular historical changes. Lehmann (2005: 7–8) distinguishes between two kinds of historical changes involving grammaticalization. In the cases of *pure grammaticalization* there is no analogical model that directs the process. In contrast, in *analogically oriented grammaticalization* processes analogy drives the course of the change. Lehmann debates the explanatory status of analogical models, because they are effective in some cases, while not in others.

In current studies there is a transformation in the way the significance and status of analogy is perceived. A discussion of different conceptions of the relationship between analogy and grammaticalization is provided in Dér (2008: 106–108). The notion of analogy appears in the literature on grammaticalization in association with the phenomenon of reanalysis. It is worth discussing both phenomena together, since current accounts of analogy also aim to defend it, primarily contra reanalysis.

Hopper and Traugott (1993: 32–62), in their book about grammaticalization, consider reanalysis and analogy to be two major mechanisms by which grammaticalization takes place and emphasize that reanalysis operates along the syntagmatic axis, while analogy operates along the paradigmatic one. Reanalysis can be defined as a reorganization of the structure of an expression which is not directly observable in its surface manifestation: it only affects the underlying (semantic, syntactic, or morphological) representation, but not the surface structure. Affected linguistic elements are reassigned to different semantic-syntactic categories.[87] In each case there is ambiguity in some contexts that allows for the interpretation of the structure as it was before and also according to a new reading (cf. Hopper and Traugott 1993: 41). In contrast, analogy modifies the surface structure. It does not lead to rule change in itself but is responsible for rule spread within the linguistic system or within the linguistic community. Hopper and Traugott (1993: 32, 57) regard reanalysis as the most important mechanism, because only reanalysis can create new grammatical structures.

The identification of reanalysis with grammaticalization also appears in the literature (Hopper and Traugott 1993: 48 mentions Meillet 1912/1921, but see also authors cited in Dér 2008: 103). According to Dér (2008: 102), grammaticalization is regarded as a type of reanalysis principally by authors within the formalist approach. In contrast, Hopper and Traugott (1993: 48–56) show that there are cases of reanalysis without grammaticalization, hence they regard them as independent processes.[88] Nowadays this approach is widely accepted and described in Dér (2005: 168) as follows: reanalysis takes place instantaneously as a part of grammaticalization, which is a gradual process.

However, Hopper and Traugott (1993) do not underestimate the role of analogy in grammaticalization. They regard it as rule generalization, and they emphasize two aspects. First, analogy makes the changes observable and provides 'evidence both for speakers of a language (and also for linguists!) that a change has taken place' (Hopper and Traugott 1993: 57). Second, in the process of grammaticalization there is a cyclical interaction of reanalysis and analogy, as Hopper and Traugott (1993: 58–59, 61) illustrate with the development of the French negator *pas*, and with the grammaticalization of the English construction *be going to*, as well.

It is worth highlighting that Hopper and Traugott (1993) maintain – as is usual in functional accounts of grammaticalization – that the main motivations for changes in grammaticalization are discourse-pragmatic factors. Reanalysis and analogy are seen as mechanisms operating in grammaticalization and not motives for change. However, grammaticalization literature has recently called into question whether it is sufficient to regard analogy as a factor at work in grammaticalization. Fischer (2010) notices that analogy has a motivating power in language change at the same time.

In contrast to Hopper and Traugott's (1993) approach, Fischer (2010: 182) argues that analogy should be not restricted to analogical extension, which is a widely recognized mechanism at work in grammaticalization, but one which also motivates language change. She regards analogy and not reanalysis as the major mechanism of linguistic change and sees analogy as a basic cognitive ability. Fischer (2010) notes that what is significant about analogy in language change is that it involves form as well as meaning and is thus suitable to serve as a framework which synthesizes formal and functional approaches. Fischer (2010) proposes an approach which focuses on both the functional factors and the formal properties of the current grammar in which grammaticalization takes place.

Fischer (2010: 187–188) claims that formal and functional approaches in grammaticalization are fundamentally different, since they focus on different forces that they believe motivate language change. Functional linguists, and especially those working within a grammaticalization framework, concentrate on semantic-pragmatic factors, and according to this focus, they deal with language use and language processing, looking at the change as gradual and driven by contextual forces. They believe that formal changes are merely the result of these processes. Analogical extension and formal reanalysis are seen in this approach as mere mechanisms and not as motivations of change. In contrast, the functional model concentrates on forces internal to the linguistic system (Fischer 2010: 185). For a detailed comparison of formal and functional approaches, see also Ladányi (1998) and Dér (2004; 2008: 58–91).

Fischer (2010: 196–197) emphasizes the interaction of functional and formal forces and tries to synthesize both approaches by emphasizing the key role of analogy and attributing a new status to it: analogy in Fischer's (2010: 193) synthesizing framework is seen as both a mechanism and as a cause. She (Fischer 2010: 194–195) maintains that, in order to describe language change properly, grammatical constraints and constraints on language use should be taken into account together. Moreover, she realizes that grammatical constraints may become arbitrary over time. She contrasts her approach with Haspelmath (1999), who believes that grammatical constraints explain why a change takes place only if we analyse them as constraints on language use. According to Fischer (2010: 196), the two motivations can be linked to two different stages of grammaticalization: at the beginning the change is local and 'can then still more easily be explained as functional or user-constrained. But when the change spreads and widens, form becomes more important than function.' It is worth noting that other authors also associate pragmatic factors with the beginnings of grammaticalization (Hopper and Traugott 1993: 87–88).

Fischer illustrates her approach with an account of the development of pragmatic markers in English. This is a case of grammaticalization, more narrowly subjectification, by which adverbial phrases which started out as adjuncts of the verb became sentence adverbials, some of which at an even later stage came to be used as pragmatic markers, which entailed not only a change in position, but also an increase in scope. In contrast to approaches that explain the scope increase by emphasizing the role of pragmatic inferencing (see Tabor and Traugott 1998), Fischer argues that it took place by analogy. Fischer assumes an intermediate stage in that the verbal phrase first appeared in a topic position at the front of the clause without a change in scope. She supposes that the use of sentence adverbials as pragmatic markers was driven by analogy, due to the semantic and structural properties that they shared with reduced modal clauses used in the same position (Fischer 2010: 199–200). The scope increase was not a contextually motivated development, but a consequence of this analogical process. Fischer attributes a more marginal role to pragmatic inferencing than assumed by Tabor and Traugott (1998), emphasizing the role of the clause-initial position and word order. In sum, Fischer concludes that:

> [T]he scope of the original VP adjunct changed presumably through its change in position and not through pragmatic inferencing of the adverbial itself. Further scope increase was presumably helped by the fact that other initial, reduced-clause adverbials with SV order already had wide scope. So on **analogy** of these reduced adverbials, which they resembled in both form and function, the original manner adjuncts could become sentential adverbs too.
>
> (Fischer 2010: 22, emphasis in the original)

Fischer (2010: 214–215), relying on all these considerations, finally calls into question the notion of grammaticalization as an independent mechanism of change. She suggests that it is an epiphenomenon, the result of smaller changes, and that stages in grammaticalization should be explained by analogy, which is seen in her model as a cause as well as a mechanism.

De Vogelaer (2010) also emphasizes the crucial role analogy plays in grammaticalization. He offers a case study of how grammaticalization pathways are determined by languages' potential for analogy. De Vogelaer deals with differences in the grammaticalization of pronouns into clitics and agreement markers in West Germanic and Romance dialects. Both West Germanic and Romance dialects have developed clitics, but they have grammaticalized further only in the Romance dialects, with the exception of some East Flemish dialects of Dutch, where the clitics have developed further, but according to a different pattern. Previous accounts of the grammaticalization of pronouns into clitics or agreement markers have considered reanalysis as the basic mechanism. In contrast, De Vogelaer (2010) explains differences in the grammaticalization pathways of clitics in the dialects studied as a consequence of the inaccessibility of certain paradigmatic extensions in West Germanic dialects, which are in contrast accessible in Romance dialects. In other words, De Vogelaer's research reveals that pathways of grammaticalization are determined by the availability of certain analogical changes, that is, the synchronic state of the grammar. The author maintains that reanalysis-based accounts cannot explain all the facts observed in grammaticalization processes, which can be better explained in an analogy-based framework which takes into account the similarity of relevant syntactic environments.

Although I have realized the present study within a grammaticalization theory framework, which belongs to the functional approach, I consider some points of Fischer's (2010) perspective useful. A combination of functional and formal factors also seems to be fruitful in the description of the grammaticalization process under study. The primary aim of the research has been to reveal pragmatic motivating forces in the semantic change of the Catalan *anar* + Inf. In contrast, Fischer proposes an account of the development of pragmatic markers in English that minimalizes the importance of pragmatic inferencing as a driving force in the process. In order to dissolve this apparent contradiction, I note that pragmatic inferencing should not necessarily have the same significant role in every grammaticalization process. I would like to highlight the following aspects of Fischer's (2010) model that seem to me useful. She tries to promote harmony between formal and functional approaches when emphasizing both formal and discourse aspects of the language-change process. In this regard her work contrasts with Hopper and Traugott's (1993) approach, where the authors, as is traditional in functional approaches to grammaticalization, put a special emphasis on the discourse-pragmatic factors motivating the process. Analogy in this model appears not as a cause, but rather as a mechanism at work in grammaticalization. A central idea in Fischer's (2010: 196–197) approach is that grammatical constraints as well as constraints related to language use should be taken into consideration for a complete account of language change. She expresses the opinion that:

> [A] change presumably starts off as a local phenomenon and can then still more easily be explained as functional or user-constrained. But when the change spreads and widens, form becomes more important than function,

again because the meaning of the more abstract structure becomes more and more bleached when this happens.

(Fischer 2010: 196)

That is why Fischer (2010: 182) thinks it is not sufficient to examine grammaticalization processes only from a functional point of view, and she emphasizes the role of the language structure in which the change is taking place. From this point of view, the framework proposed in Fischer (2010) seems to be very useful in revealing the history of the Catalan GO-construction, too. This grammaticalization process, which does not fit any hitherto detected grammaticalization cline, reveals what an important role the particular grammatical structure of a language can play in these processes. In this sense, the instance of grammaticalization presented in this book allows for an interpretation in terms of Fischer's model. The history of the Catalan *anar* + Inf construction is made up of small steps that in themselves parallel changes detected in other languages around the world. The big picture, however, develops under the influence of the grammar of Catalan, evolving in this way a particular pattern of change. Analogical processes that have been assumed by Juge (2006) (see below in Section 3.6.4.4) to play a crucial role at a certain point of the change, could be at work due to particular properties of the grammatical system of Catalan. The reanalysis of the construction as a periphrasis with present tense auxiliaries has been possible due to the particular grammatical system of Catalan. At the same time, I do not agree with Fischer's (2010) claim that grammaticalization is an epiphenomenon rather than an independent mechanism of change. Nevertheless, I consider the focus she puts on both formal and functional aspects of language change processes to be useful.

In summary, the notions of frequency and analogy play a central role in historical linguistics argumentation. Frequency-based argumentation makes use of plausible statements which capture characteristics of the linguistic structures in quantitative terms, while analogy-based argumentation refers to analogical processes supposed to be at work at a given linguistic level. Since they are widely accepted methods and explanations in this field, statements from research relying on them can be assigned a high plausibility value.

3.6.4 Deciding between competing hypotheses

3.6.4.1 Introduction: rival solutions in the starting context

Our starting p-context, presented in the object scientific chapter of this book, contains statements concerning the grammaticalization of *anar* + Inf which are made plausible by some sources and implausible by others, i.e. they yield a p-inconsistent, informationally overdetermined p-context (cf. Section 3.6.2.4). The hypotheses responsible for this inconsistency come from the relevant literature, and among them those concerning some morphological aspects of the grammaticalization process under consideration are especially significant. The problem consists in the following. Although auxiliary forms of the perfective past *anar* + Inf of present-day Catalan formally seem to be in the present tense, medieval data attest an early alternation between present and preterit auxiliary forms. The main difference between accounts of the morphological development of the *anar* 'go' + Inf is how they judge the status of early occurrences with the verb 'go' in the present tense. In the present section I examine three accounts of the morphological development of the *anar* 'go' + Inf construction, namely, those of Colon (1959/1978, 1976/1978), Detges (2004), and Juge (2006) from a methodological point of view, in order to assign a plausibility value to their statements about morphological aspects of the process. These authors are representatives of the three most widespread hypotheses concerning the history of the Catalan GO-past, presented in Section 2.3.[89] The aim of the comparison is to decide between rival statements in order to eliminate the p-inconsistency of the starting context. I do not consider here the deixis hypothesis, because its presence in the starting p-context does not contribute to the inconsistency. The reason is that it does not deal with the contexts and morphological aspects of the process, thus it can be reconciled with any of the other hypotheses and can complete them. Consequently, it cannot be treated as a competing hypothesis, in other words, we do not have to accept or reject it now. Let us begin the comparison with the presentation of Colon's (1959/1978, 1976/1978) arguments.

3.6.4.2 Colon (1959/1978, 1976/1978)

Concerning the question of whether the origin of the Catalan GO-past is the periphrasis formed with the present or past tense auxiliary, Colon (1959/1978, 1976/1978) provides a 'narrative present account', which derives the current Catalan *anar* + Inf construction from a historical present usage (cf. also Badia i Margarit 1981). From a semantic point of view, such an analysis is primarily based on the medieval occurrences of *anar* + Inf with a present tense auxiliary. Colon does not deny the existence of preterit auxiliary occurrences in the historical texts, but quite simply renders them invisible to the semantic and pragmatic analyses. According to his analysis, the current meaning of *anar* + Inf can be derived from the present auxiliary variant. Colon (1976/1978: 161) argues that the reason for the loss of preterit forms in the auxiliary was a historical present use which pushed the preterit forms of the auxiliary into the background. He (Colon 1959/1978: 128–129; 1976/1978: 161) explains this in the following way: the use of the GO-periphrasis and the historical present fulfil similar functions, inasmuch as they both make the description more vivid and attract the hearers' interest. Since the construction *anar* + Inf serves to animate the description, it is perfectly understandable that its auxiliary is conjugated in the historical present instead of a preterit, because its role is the same, and both 'unite their virtues'.

In order to illustrate the claim that the construction *anar* + Inf serves to animate the description, Colon (1959/1978: 127–128) mentions an interesting example which shows the auxiliary in the present tense. This example of *anar* + Inf appears in a general description in the present tense in a Catalan text, which has a parallel text in Provençal. In the corresponding Provençal version, a simple present appears where Catalan uses the GO-periphrasis with a present tense auxiliary.[90] Colon attributes this use to the function of animating the narration. This occurrence does not correspond to the formally similar occurrences which show the periphrasis with a present tense auxiliary in a preterit context, because it appears in a general description in the present tense, i.e. the context is different. This piece of data from a direct source certainly makes Colon's hypothesis about the early function of the GO-construction (animating the narration) plausible. At the same time, however, it makes the following claim plausible, too: in the earliest examples, the auxiliary should appear in a tense required by the context (cf. Colon 1959/1978: 128; 1976/1978: 159). Consequently, it can be supposed that in the earliest examples the auxiliary of the periphrasis had to be conjugated in the preterit in a preterit context.

Concerning the later morphological development of the construction, Colon (1976/1978: 169) only refers to the appearance of the *var-* forms. The *var-* auxiliaries (cf. Table 1) were created by analogy, modelled on the regular preterit forms of first conjugation verbs. Colon suggests that the formal similarity of *var-* auxiliaries to regular perfective past forms attests to the temporal identity of the construction. That is, he supposes that the construction had already been grammaticalized as a perfective past tense by that time (about the 14th century).

Let us consider a summary of the development of *anar* + Inf in Colon's (1959/1978, 1976/1978) account, step by step:

1. *Anar* + Inf is a periphrasis with a motion verb conjugated as the context requires, used in order to animate the narration.
2. The use of auxiliaries in the historical present extends in a preterit context, and this historical present usage displaces the preterit auxiliary forms. The reason is that the historical present and the GO-periphrasis have the same function and 'join their virtues.'
3. The preterit forms of the auxiliary disappear, relegated to the background by the narrative present usage. The loss of the simple preterit tense also plays a role in this process.
4. The periphrasis – with the GO-verb in present – acquires a past tense value.
5. As a consequence, the analogical *var-* forms appear, demonstrating the temporal identity of the construction.

The statements 1–3 are explicit claims in Colon (1959/1978, 1976/1978). However, because the plausibility of a conclusion does not only depend on the plausibility values of the premises but is also influenced by the plausibility of the latent background assumptions, we have to reconstruct them before considering the plausibility of Colon's hypotheses. Latent background assumptions are statements which are also needed to secure the plausibility of the conclusion in certain cases (cf. Kertész and Rákosi 2012: 89–91). The following latent background assumptions can be reconstructed at the crucial second and third points of Colon's argumentation. First, he supposes that two things which have the same function can fulfil this function better together. However, he does not support this claim by independent evidence. Second, he supposes that the present auxiliary forms in medieval texts are representatives of the historical present. Third, he supposes that the historical present has the function of animating narration (cf. Colon 1959/1978: 129; 1976/1978: 161). These latter two latent background assumptions show that Colon's argumentation concerning the morphological development of the *anar* + Inf periphrasis is based on an assumed historical present use. This view is, however, not properly established: Colon does not express explicitly what he means by 'historical present' and does not support these assumptions by other data. Because none of them is supported properly, low plausibility values can be assigned to these latent background assumptions and, as a consequence, the plausibility of the hypothesis decreases too.

As to the shift between the third and fourth points of the development, according to Colon (1959/1978, 1976/1978: 156) we cannot examine the beginnings of the history of the Catalan *anar* + Inf construction, because they already appear with a past tense value in the earliest texts. In contrast with Colon's claim, on the basis of corpus data it can be claimed that in the earliest text the range of contexts in which the periphrasis can appear is very limited.[91] The grammaticalization of the construction has not yet concluded, and if there is a past tense value, it is not yet a grammatical meaning. The fifth claim can be assigned a high plausibility value on the basis of data from several different language systems (see Juge's (2006: 316) argumentation, discussed under 3.6.4.4).

Colon's analysis is based on a wide range of occurrences of GO + Inf, not only from Old Catalan, but also from other Romance languages. However, his account also contains an inconsistency between two of his statements, namely, (a) that the periphrasis has a past tense value, and (b) that it deals with a historical present use. This inconsistency results from the circumstance that both claims are based on the same sources of data, that is, the same historical texts, which represent the same language state. It is obvious that the two statements, inasmuch as they concern the same language state, cannot be true simultaneously. If the first statement is maintainable, the GO-verb should already function as a grammatical marker, or at least be grammaticalized to some degree. However, this would contradict the statement that we are dealing with a historical present usage, because in this case the GO-verb should really refer to motion. Consequently, the simultaneous presence of both statements in the argumentation would yield a p-inconsistent p-context. However, there is another theoretical possibility: we can suppose that Colon intends the two statements to refer to two different language states. In this sense, the second one concerns an earlier period of the history of Catalan, from which, as Colon also points out, we do not have documents. Given the lack of direct sources from this language period, it is not possible to support this claim properly. Consequently, it receives a low plausibility value, which decreases the plausibility of Colon's account.

3.6.4.3 Detges (2004)

Detges also claims that the present auxiliary variants of *anar* + Inf are representatives of a historical present usage. Detges's (2004) argumentation is also similar to that of Colon in that he attributes to the present tense auxiliary variant two effects: dynamism (due to the verb 'go') and presentification (due to the historical present). He claims that the tense of the auxiliary is optional, although the present-tense variant was used more frequently at all times. He also maintains that the verb *anar* in the present auxiliary periphrasis was conventionalized as a foreground-marker due to its high frequency. Let us reconstruct Detges's (2004: 218) argumentation concerning the present-past alternation in the auxiliary of *anar* + Inf in more detail.

1. The narrative genres of medieval Western Romance show an unsystematic present tense–past tense alternation, especially in earlier texts.
2. The historical present is dispensable for the 'effect' of the periphrasis; the tense of the auxiliary is optional.
3. The version which has the auxiliary in the present is the 'stronger alternative'. The present tense realization combines two effects, namely, the 'dynamification effect' and, due to the historical present, 'the presentification effect'.
4. The present tense variant was used more frequently at all times.
5. As a consequence of its higher frequency, the present tense variant came to be conventionalized as the only possible form.

The problem with the first claim is that although it may be maintainable, it is hard to see whether it is relevant enough to include in the argumentation. Although it is possible that the formation of the past value of this Catalan construction took place in the context of such an incoherent verbal tense use, Detges does not provide examples taken from historical documents with this characteristic. Moreover, this claim does not hold for the medieval Catalan chronicles, which contain the earliest documented occurrences of the periphrasis under study. Detges himself presents examples[92] which testify to a coherent use of preterit tenses in narrative, only interrupted by occurrences of the GO-periphrasis with an auxiliary in the present, but with no other present tense forms nearby. The relevant example, taken from Colon (1976/1978: 135–136) and provided in Detges (2004: 217) is a passage from Desclot's Chronicle. It contains 14 verbal forms, of which 13 are in the past tense, and the only one in the present is the auxiliary of an occurrence of the GO-periphrasis. As another example he (Detges 2004: 221)[93] presents a passage from Muntaner's Chronicle (again taken from Colon 1976/1978: 169). This contains 23 verbal forms. Two of them appear in a direct quotation and two others in the expression addressing the public *Què us diré?* 'What shall I tell you?' The rest of them belong to the real narrative part of the text, and five of them are representatives of the *anar* + Inf periphrasis. Of these five occurrences only one has the GO-verb in the preterit, the other four are occurrences in the present tense. However, these present GO-verbs are the only present forms with a past tense reference in the text. That is, there are no other present forms which could be representatives of the historical present: the coherent verbal tense use in this narrative is only interrupted by the present variants of *anar* + Inf. The illustrative examples do not show the unsystematic alternation of verbal tenses assumed by Detges. Nevertheless, medieval Catalan narrative texts are not characterized by such an alternation, but a coherent use of tenses; the periphrasis is the only verbal form conjugated in the present and appears only in a limited range of contexts. Consequently, this is not a felicitous argument, or at least it should be mentioned that the periphrasis also appears in texts whose verbal tense use is systematic and coherent. The present tense forms in this type of texts should be explained, too. That is, the p-context of Detges's investigation is p-inconsistent in the sense that the statement concerning the 'unsystematic present tense–past tense alternation' is claimed to be plausible, while direct sources make its negation plausible.

As for the second claim, Detges does not explain why, if this is the case, the historical present would appear at all, and why only in the case of this periphrasis. At this point a latent background assumption can be reconstructed, namely, that the present auxiliary occurrences of *anar* + Inf are representatives of a historical present usage. However, since he operates with an intuitive notion of the historical present, this statement is not properly supported. Consequently, it has a low plausibility value, which weakens the plausibility of his hypothesis as well.

Detges bases the third claim, again, on the latent background assumption that the present auxiliary versions are representatives of a historical present usage. He argues that they are stronger versions, since they combine two effects. But how can we know what 'effect' the periphrasis had in the Middle Ages? We should support such a claim with evidence accepted in historical linguistics. If a stronger and a weaker version of the periphrasis actually existed in the medieval language, such a difference should be reflected in the documented uses. However, contexts in historical documents do not reflect a 'weaker version–stronger version' opposition: in early texts the present and the past tense auxiliary versions appear in the same types of contexts and with the same infinitives. Consequently, we do not have supporting evidence for the claim that the present auxiliary variant was stronger. This statement is only based on the author's subjective judgment, and our sources (i.e. historical documents) make it implausible by supporting its negation. Once again, the extension of the p-context of Detges's analysis with new data results in a p-inconsistency.

The fourth claim concerns the present tense variants of the GO-construction, which according to Detges (2004) was used more frequently at all times than preterit auxiliary occurrences. Detges does not support the higher frequency of the present auxiliary version with (quantitative) data. Thus, the reader can misinterpret this statement, believing that statistics have been used in the investigation, which is not the case. Moreover, statistical data suggest the opposite hypothesis, according to which the preterit auxiliary variant of *anar* + Inf was more frequent at the beginning and was displaced by the present auxiliary forms only later (cf. Colon 1976/1978: 159–160, and data of the present study shown in Tables 3–4). In other words, the extension of the p-context of his investigation with new information originating from these statistics yields a p-inconsistency. We have to decide, then, which one of the two simultaneously plausible statements to accept: statement 4 or its negation.

Finally, Detges attributes the loss of the preterit forms to the higher frequency of present auxiliary variants and their consequent conventionalization. Although frequency can have this effect, Detges does not support this claim with a frequency analysis of historical occurrences, neither does he take into account results presented in previous literature, which show the opposite: at the beginning the preterit auxiliary occurrences were more frequent. Consequently, on the basis of information from relevant literature we have to judge this statement to be implausible.

To sum up, Detges's argumentation does not rely sufficiently on data whose plausibility value originates from direct sources, and in his argumentation there appear statements which are not presented as results of usually accepted historical linguistics argumentation methods. At the same time, when he uses accepted argumentation methods (e.g. based on frequency), his statements are either not supported by data whose plausibility value originates from some direct source, or further sources render them implausible.

In addition, a textual contradiction appears, confusing the reader. Detges (2004: 213) translates the Old French form *(il) va dire* in the following way: 'not simply "he starts to say", but more specifically "all of a sudden, he says" (Gougenheim 1929/1971: 96)'. And elsewhere he says (Detges 2004: 217): '*il va dire* "suddenly, he said" (Gougenheim 1929/1971: 96–97)', this time with a translation in the preterit. We do not know whether this contradiction between two different translations with present and past tenses comes from Gougenheim or it is Detges's own mistake. In any case, it is very confusing, as the question of tense is a crucial one in the development of this construction.

In Colon's (1959/1978, 1976/1978) and Detges's (2004) accounts presented above, the morphological analysis of the development of the Catalan GO-past amounts only to the 'historical present explanation'. However, it is in itself not properly established. Consequently, the p-context of the research containing these approaches should be extended with new information about historical present usage. In addition, these two analyses need to be complemented with more detailed morphological and pragmatic analyses, because the plausibility of a statement which is in accordance with analyses of other linguistic levels is higher. As Kertész and Rákosi (2012: 76) point out, a statement may be also supported by several different sources. In such cases, its plausibility value is higher on the basis of all sources together than its plausibility value on the basis of any of the sources alone.

3.6.4.4 Juge (2006)

The third account of the grammaticalization of the Catalan GO-past presented here is Juge's (2006) analysis, which examines the morphological aspect of the process more thoroughly than the previous accounts. Juge (2006: 314) characterizes the history of *anar* 'go' + Inf in the following way: '[T]his seemingly puzzling development results from the interaction of a number of commonly found diachronic processes with language-specific constructional patterns'. Unlike the other two accounts, Juge takes the past auxiliary forms as the starting point in his analysis. As a first step, he analyses the morphology of the auxiliary and compares it with the paradigm of the full lexical verb *anar* 'go'. The main difference consists in the 1Pl and 2Pl forms, which showed a present–preterit syncretism in Old Catalan.

Juge (2006) claims that present tense forms in the current paradigm result from analogy and that this process only took place later, when the past meaning of the whole construction had already consolidated. The syncretism of a very current form of *anar* 'go' (*anam* 'we go/we are going/we went', and the second person plural *anats*) made it possible to reanalyse the periphrasis as a construction with a present tense auxiliary. As a consequence, the use of the present extended to the rest of the paradigm of the auxiliary. Let us observe his methods and argumentation in more detail.

1. At the beginning, *anar* + Inf is a periphrasis with a preterit auxiliary in narrative contexts.
2. A semantic change can already be detected in some preterit auxiliary forms. That is, *anar* + Inf with a preterit auxiliary becomes a past tense.

3. A reinterpretation of the following type takes place: *anar* (preterit) + Inf → *anar* (present) + Inf. To put it differently, language users reinterpret the periphrasis as a construction with a present tense auxiliary. Frequency and analogy trigger this process. The result is that *anar* + Inf is now a periphrasis with a present tense auxiliary.

4. Later, new forms are created by analogy with regular preterit forms. Juge (2006: 316) argues that other forms are analogically created from the root *va-*, since the form *va* '(he) goes' is both monosyllabic and third person singular.

5. The present auxiliary occurrences do not show the characteristics of a historical present use.

Juge (2006) supports the first claim by frequency analysis; consequently, a high plausibility value can be assigned to it. Juge (2006: 319) presents corpus data in support of the second claim, too: in his corpus, 32 of the 157 (20.4%) instances of *anà* 'he went' 'are plausibly cases of the periphrastic preterit' and, similarly, with *anaren* 'they went' there are 164 instances in all, with 40 (24.4%) of these 'being good candidates for the periphrasis'. The third point of the development described in Juge (2006) is crucial. At this point, language users presumably reinterpret the periphrasis as a construction with a present tense auxiliary. In this reinterpretation, Juge (2006: 320) attributes a crucial role to the ambiguity between present and preterit found in the first and second plural indicative (cf. Table 2 in 2.3.6). Because present and preterit are formally identical in the first and second plural indicative, language users reinterpret the construction with a preterit auxiliary as one with a present auxiliary. Juge (2006: 320) claims that 'the influence is not directly on the other forms in the paradigm; rather it is on the interpretation of the structure of the construction as a whole', that is, at this point Juge assumes an analogy with other constructions which have the auxiliary in the present tense.

The analogy-based and frequency-based arguments are connected here. On the one hand, according to Juge (2006: 321), 'one possible reason for the reanalysis of forms such as *anam* as present in this construction is the tendency for finite verbs participating in periphrases to be morphologically present tense'. Juge's argumentation, relying on analogy with formally similar constructions, takes into account the whole verbal system. His other argument is based on cross-linguistically recurrent tendencies: 'there is a strong cross-linguistic tendency for auxiliary marking to be in the present, although this is certainly not an absolute universal' (Juge 2006: 322). That is, the reinterpretation of the construction *anar* + Inf as a construction with a present tense auxiliary is triggered by the existence of other periphrases with a past reference, which have a present or at least a non-perfective auxiliary. On the other hand, Juge claims that it is frequency which triggers this process. As he (Juge 2006: 320) points out, 'such forms, especially the 1Pl, are abnormally frequent because they are first person narratives'. Table 27 shows the data provided by Juge concerning the frequency of different forms of *anar*.

Table 27. 'Frequency of first person plural and other forms in Old Catalan' Juge (2006: 321)

	Preterit		Present indicative	
	Form	**Frequency**	**Form**	**Frequency**
1Sg			*vag*	1
2Sg			*vas*	1
3Sg	*anà*	41	*va*	13
1Pl	**anam**	209	*anam*	(209)
3Pl	*anaren*	71	*van*	10

In Juge's corpus, 209 of 346 occurrences represent the 1Pl form *anam* 'we go/ we are going/we went', which is a relatively high frequency. The frequency-based argumentation at this point consists in claiming that the reinterpretation of *anar* + Inf as a periphrasis with a present tense auxiliary is realized due to the high frequency of the 1Pl form *anam*. This form is ambiguous between past and present tense and allows an association with the present tense paradigm of the verb *anar*.

Two problems should be mentioned, which can decrease the plausibility of this statement. First, in other cases of analogical change, the basicness of third person is emphasized. Second, it is questionable whether the frequency will remain the same when investigating it in a larger corpus. As to the first problem, the extremely high frequency of this no-third person form can explain its central role on this particular occasion, because it can be considered a basic form. Juge's argumentation at this point is in accordance with Kuryłowicz's second 'law' of analogy: 'analogical developments follow the direction "basic form" → "derived form", where the relationship between basic and derived forms is a consequence of their spheres of usage' (Hock 1991: 212). Also: 'of any given set of forms or morphological classes, the one which has a greater sphere of usage is more "basic" than the others' (Hock 1991: 214). That is, the 1Pl form *anam*, due to its high frequency, can be considered a basic form. As to the second problem, at the moment, it is a plausibility weakening factor, and needs more investigation.

The fourth claim, concerning the forms analogically created on the root *va-*, can be supported by cross-linguistic data, since the basicness of the third person can be found in many languages and often has diachronic consequences (cf. Hock 1991: 220). Third persons tend to be more basic in analogical change than other forms of the verb (Hock 2003/2005: 446; cf. also Kuryłowicz's 'sphere of usage' provision in Hock 1991: 220–222). Some forms created on the root *va-* resemble regular preterit forms like *cantares* 'you sang'. These are also given an explanation in Juge (2006: 316). The marking of the periphrasis with a preterit meaning by a preterit marker also has a parallel in another language. As Juge (2006: 316) points out, in Russian there is a parallel for such a 'double' marking where the tense of the auxiliary coincides with that of the construction as a whole.

Finally, the last, fifth statement, according to which the present auxiliary forms do not show the characteristics of a historical present use, is crucial in the argumentation in Juge (2006). Juge (2006: 319) supports this claim by another one: all the present tense auxiliary occurrences found in his corpus are representatives of the periphrasis (that is, they do not retain the lexical meaning 'go' of *anar*). As he emphasizes: 'there is NOT A SINGLE EXAMPLE of *va* or *van* used non-periphrastically with a past value' and the present tense variants appear 'with no other present tense forms nearby' (Juge 2006: 319, emphasis in the original). Similar to the other two authors, Juge (2006) does not explain exactly what the historical present is. Although his claims concerning this issue are intuitively acceptable and seem to provide a correct description of the phenomenon, with an extension of the p-context with new, relevant information, the plausibility of his statements can also be modified. In a later work, Juge (2008) provides a notion of historical present and a detailed argumentation against the hypothesis that there is a historical present use in these texts.

To sum up, Juge's (2006) argumentation takes into consideration the whole verbal system and cross-linguistic tendencies, too. He often makes typological comparisons and relies on examples and parallels from other languages. Juge bases his claims on data originating from a historical corpus and provides an exhaustive morphological analysis of occurrences. He explains the morphological development of the construction from the beginnings until the modern paradigm with argumentation methods accepted in historical linguistics. References to analogical processes and frequency are widely used in his argumentation, yielding a convincing analysis.

3.6.5 Extension of the starting p-context and coordination of the extended p-context

Our starting p-context has been problematic, because it has been informationally overdetermined: the simultaneous presence of certain statements and their negations in it has resulted in a p-inconsistency. Namely, in our context there have been hypotheses assuming a historical present use in present tense occurrences of *anar* + Inf, and also hypotheses which reject the idea that they were historical present forms. The main difference is that the two analyses posit the semantic change at different points of the development. In the first analysis it is the periphrasis with the present auxiliary which undergoes semantic change, while in the second account it is the periphrasis with the preterit auxiliary that does so. Similarly, the spread of the present auxiliary forms is accounted for differently in both analyses. In the first, it is the result of an extension of the historical present usage (which, by the way, very strangely only affects the forms of this periphrasis). In the second analysis, however, they are the result of a reinterpretation, triggered by the syncretism in one of the most frequent forms of the paradigm. According to Juge's (2006) theory, it was the change in meaning which triggered formal change. This happens frequently in analogical changes: the users of a language give similar forms to semantically similar linguistic units (cf. Hock 1991: 167).

The extension of the p-context with new information concerning the historical present (as discussed in 2.6.7) helps us to eliminate one of the p-inconsistencies present in the starting p-context, namely, that which exists between the statement that there is a historical present use in the historical documents under study (considered as plausible in Colon's and Detges's accounts) and its negation (made plausible by Juge's approach). On the basis of methodological considerations and new information concerning historical present usage, we should modify the plausibility values of these statements: the statement that it is about a historical present use should be evaluated as implausible, while the plausibility of its negation should be strengthened. Accordingly, present occurrences of *anar* + Inf in medieval texts should not be considered representatives of a historical present usage but rather forms which, in comparison with the preterit variants, show a more advanced stage of grammaticalization.

3.6.6 Modification of the p-context and comparison of the rival solutions

At this point of the investigation, we have to compare the rival solutions, decide which one to accept, and, with the modification of the p-context, eliminate the p-inconsistency. In historical linguistics, pieces of information concerning cross-linguistically recurrent tendencies in linguistic change often help us to decide between competing hypotheses, since they can be used as data and be included in the inferential process, and in this way they can influence the plausibility value of certain statements. Let us discuss how considerations about cross-linguistically attested tendencies in analogy and in semantic change in grammaticalization appear in historical linguistic argumentation. Various cross-linguistic tendencies have been detected concerning the way in which meanings change in the languages of the world (cf. Bybee et al. 1994; Heine and Kuteva 2002 among others), but also in the case of analogy we find some tendencies concerning the way in which analogical change usually occurs. Argumentations in linguistics which make use of cross-linguistically recurrent tendencies also rely on plausible inferences; because linguistic universals cannot be regarded as certainly true statements, they can be only plausible. Raising hypotheses about universal characteristics of languages starts from the investigation of a limited number of languages, when the presence of the characteristics at issue is checked, and, in a second step, an inductive inference is applied, where 'the set of the premise-candidates has to be supplemented by the background assumption that the cases not examined also possess the characteristics that could be found in the investigated ones; the conclusion states the presence of these characteristics as a general rule' (Kertész and Rákosi 2012: 90). Kertész and Rákosi examine various examples of plausible argumentation which concern linguistic universals (see especially example 12 in Kertész and Rákosi 2012; for the testing of potential linguistic universals, see also examples 4(1), 9(3), 19, 20(2) in Kertész and Rákosi 2012).

Statements concerning linguistic universals can function as data in historical linguistics as well and appear in the set of premises of inferences which aim to judge the plausibility of some hypothesis of the theory (cf. Heine 2003/2005: 580, 585f.). In relation to natural directionalities in analogical change, Hock (1991: 210) also points out that 'they might also be helpful in choosing between alternative possible analyses. Everything else being equal, we would select that analysis which better agrees with what is known to be more natural.' That is, statements concerning cross-linguistically recurrent tendencies of change can enter the argumentation process and influence the plausibility value of statements.

Since I have already presented above how tendencies in analogical change appear in historical linguistic argumentation (cf. Juge's analysis in 3.6.4.4), in the present section I only examine how considerations about cross-linguistic tendencies in semantic change can help us to choose between competing accounts of a historical phenomenon. The question of how the development discussed in the present volume fits cross-linguistically attested tendencies of semantic change is twofold. First, we can examine which lexemes past tense markers usually develop from, and, second, what semantic change lexemes meaning 'go' usually undergo. Regarding the first question, the development of the verb 'go' conjugated in the present tense in combination with an infinitive into a past tense does not fit any cross-linguistically frequent tendency (cf. Heine and Kuteva 2002). As to the second question, the lexeme 'go' can suffer different semantic developments in different constructions. Let us consider the following statements based on Heine and Kuteva (2002).

a. The purposive construction 'GO-verb conjugated in the present tense + infinitive' tends to develop into a future tense in the world's languages.
b. There is no language in which the purposive construction 'GO-verb conjugated in the present tense + infinitive' has developed into a past tense.

We can use these general claims about the semantic development of lexemes meaning 'go' in assigning plausibility values to the following two statements:

1. The Catalan GO-past comes from the purposive construction 'GO-verb conjugated in the present tense + infinitive'.
2. The Catalan GO-past does not come from the purposive construction 'GO-verb conjugated in the present tense + infinitive'.

In the case study presented in Chapter 2, our starting context contained several hypotheses concerning the morphological development of the Catalan *anar* + Inf construction. Some of them made statement (1) plausible, while others statement (2). Consequently, the context has been inconsistent. We have extended the starting context with new information and re-evaluated the plausibility of the three main hypotheses. Now, we have to modify the new context and decide between the rival hypotheses. I want to show how taking into account cross-linguistically recurrent tendencies helps us to do that. A high plausibility value can be assigned to the statements in (a) and (b), because they are supported by several case studies in historical linguistics. Relying on (a) and (b), we can assign a high plausibility value to statement (2), because it is in accordance with general tendencies presented in (a) and (b), while statement (1) can be considered implausible on the basis of (a) and (b). In the present form, however, statement (2) is too general. As presented above in Section 3.6.4.4, Juge (2006) claims that (3) *the Catalan GO-past comes from the purposive construction 'GO-verb conjugated in the preterit + infinitive'*. This statement can be regarded as a special case of (2). Moreover, this latter hypothesis can be supported by cross-linguistic analogy with the Tucano language, which shows a similar development. When Bybee et al. (1994) examine possible paths for the formation of perfectives and past tenses, they mention the completive aspect as one of the possible sources. They define the meaning of completives as 'to do something thoroughly and completely', but they point out further semantic nuances and other uses completives may have, among them that 'the action is reported with some emphasis or surprise value'. They report the 'go' completive in the Tucano language as one which resembles 'the English "went and did it" construction: *He went and told her the whole story …* which is usually used in the past to emphasize the deliberateness and finality of an action' (Bybee et al. 1994: 57). The case of the Catalan *anar* + Inf is very similar: it also seems to refer to a completed action, and, at the same time, it expresses emphasis. Thus, the development of this construction can be supposed to follow the path described in Bybee et al. (1994: 104) as 'that of resultative or completive leading to anterior and then to perfective or simple past'. That is, completives are a possible source for perfective aspect and then past markers, and the history of the Catalan *anar* + Inf seems to instantiate this process. Such an account is also in accordance with a historical pragmatic analysis (see Nagy C. 2010).

The initial question in the present case study is as follows: from the two documented variants of the *anar* + Inf construction, namely, (a) *anar* 'go/auxiliary' conjugated in the present + Inf and (b) *anar* 'go/auxiliary' conjugated in the perfective past + Inf, which one should be considered the ancestor of the perfective past of modern Catalan? Now, we can answer that it is plausible that it is the second one. This decision favours Juge's (2006) account and makes Colon's (1959/1978,

1976/1978) and Detges's (2004) claims concerning the morphological aspect of the development of *anar* 'go' + Inf implausible. Juge's (2006) hypothesis deserves a high plausibility value primarily due to its complexity: his argumentation relies widely on frequency, analogy, and cross-linguistic similarities, and the development he describes fits general tendencies of semantic and analogical change.

3.6.7 Conclusions

In Section 3.6 I have discussed how metalinguistic considerations can help the decision between competing hypotheses in linguistic research. I have presented three approaches concerning the morphological aspect of the history of the Catalan *anar* 'go' + Inf construction: those provided in Colon (1959/1978, 1976/1978), in Detges (2004), and in Juge (2006). I would like to highlight again that I have not presented these approaches in their entirety. I have only taken into consideration claims that concern the medieval present–past alternation in uses of the Catalan GO-construction. Even if some imperfections in their argumentation can be detected, these authors all provide very valuable findings on different aspects of the history of this Catalan GO-periphrasis. Colon (1959/1978, 1976/1978), in particular, provides a great amount of information about the medieval use of this periphrasis, which should not be less appreciated even if we reject the historical present hypothesis. The three accounts considered have formed a p-context in which some statements and their negations concerning the sub-question discussed in the case study have been plausible simultaneously, i.e. our starting p-context has been p-inconsistent. This p-inconsistency has been eliminated by the extension of the starting p-context and, then, the coordination of the extended p-context. We have re-evaluated and compared plausibility values of different statements present in the p-context, also taking into account some widely accepted argumentation methods in historical linguistics.

As regards the sub-question discussed in the present case study, the main problem with the historical present accounts has perhaps been exactly what Fischer describes in the following way:

> There is a natural tendency to interpret an older construction very much from the point of view of the modern system. This happens especially when the form of the construction has remained more or less the same.
>
> (Fischer 2007: 18)

It seems that the formal similarity between forms of the current paradigm of *anar* + Inf (which have the auxiliary in the present) and the medieval forms of *anar* + Inf conjugated in the present tense might have led to a superficial analysis in this respect and distracted attention from the preterit occurrences of the medieval *anar* + Inf. This has yielded the omission of a more thorough morphological analysis, which has appeared only later, in Juge (2006). The development of the Catalan *anar* 'go' + Inf construction has found a more satisfactory explanation under this latter proposal.

It is important to note that the emphasis frequency counting has been given in the present volume does not mean that the corpus method and quantitative data are the only ones to be used in historical linguistics. The methods discussed in the present book do not cover all the methodological tools used in historical linguistics. In their introductory study to a volume of selected articles on historical pragmatics methodology, Fitzmaurice and Taavitsainen (2007: 17, 18) point out that 'it is not clear that corpus methodology is applicable to purely pragmatic research questions' and that 'corpus-based methods can be barriers to the investigation of pragmatic phenomena like conversational implicatures, which are neither routinely nor conventionally realized in lexical grammatical expressions'. To put it differently, in historical pragmatics studies, qualitative methods also have a great significance (cf. Navarro 2008), as also exemplified in several places in the current volume. However, discussion of the limits that the corpus method sets for the research of some, primarily historical pragmatics phenomena is beyond the scope of our subject.

Grammaticalization of the Catalan *anar* + Inf construction

4.1 A new hypothesis on the grammaticalization of the Catalan *anar* + Inf construction

At this point of the research, we have to answer research question (ix):

> ix. Why did the Catalan construction follow a different evolutionary path from that followed by similar motion verb constructions in other languages around the world?

First, let us discuss explanations provided in the previous literature. According to Pérez Saldanya (1996) and Pérez Saldanya and Hualde (2003: 56–59) one of the reasons why the Catalan *anar* + Inf grammaticalized into a past tense was the disappearance of the simple past. Several morphological problems with the simple past induced the speakers to use the GO-construction with greater frequency, and a homophonous construction with future meaning did not become established in Catalan.[1] Some authors also emphasize the role of the period known as the Decadence (from the 16th to the18th century), when tendencies originating from colloquial and vulgar language use could spread in the absence of a literary norm (see Colon 1976/1978; Detges 2004). In contrast, Juge (2006, 2008) regards the success of the periphrasis with *anar* in Catalan as a consequence of the coexistence of various – principally morphological – idiosyncratic factors.

Nevertheless, on the basis of considerations presented in this volume, we can hypothesize that the evolution of the Catalan GO-construction also fits a cross-linguistically recurrent tendency, inasmuch as it followed the pathway presented above until the 'intention' stage, which I repeat here for convenience:

'the motion path': motion with the aim of doing something > intention > (future)

(Bybee 2002: 181).

I have compared above occurrences of 'go' + Inf with functionally similar medieval periphrases (THINK OF + Inf and BEGIN TO + Inf). The comparison of these three medieval periphrases in Catalan and Spanish has revealed a similarity between the uses of *anar / ir (a)* 'go' + Inf and *pensar (de)* 'think of' + Inf, and the difference of both from the use of *començar (a/de)* 'begin to' + Inf. I claim that the first stage of the history of the Catalan periphrastic perfective past coincides with the first stage of the above-mentioned future tenses. Analysis of historical occurrences seems to confirm the hypothesis that both the Catalan *anar* 'go'+ Inf and the Spanish *ir (a)* 'go (to)' + Inf had a stage in their history when the meaning component 'intention' was salient within the semantic structure of the verb 'go'. These periphrases probably fulfilled a pragmatic function at the beginning of their grammaticalization, which consisted in the dynamization of the storytelling, by referring to intentionality, i.e. the subject's volitional involvement in the action.

Additional evidence comes from the finding that there is another pragmatic function of the GO-construction which can be derived from the meaning 'intention'. In the Catalan Example (97), the periphrasis is used in order to give an emotional force to the utterance.

(97)

*Barons, ¿no sabets vosaltres que el rei de França és nostre cosí germà, e misser Carles atretal? Doncs, ¿con podets consellar que jo **vaja pendre** misser Carles?*

'Barons, do you not know that the King of France is Our cousin-german, and Sir Charles as well? Then, how can you advise me **to seize/(to go and size) (= to go to seize)** Sir Charles?'

(Munt II 65,8–10)

The second utterance in (97), without any doubt, can also refer to a motion. In order to seize Charles, the king would have to move in space. The speech act verb *consellar* 'advise' signals a directive speech act. But the advice likely refers to taking and not to moving. The act of taking Charles seems to be more important and the verbal form in the subjunctive *vaja* 'me to go' seems to have an emphasizing function. The utterance apparently has the semantic nuance of indignation. This reading is also confirmed by the question form *¿con podets ...?* 'how can you ...?', which is a conventional form used to express indignation. The content of the preceding utterance, which is in close relationship with the utterance containing the GO-construction, describes the circumstance which explains the king's indignation. Both clauses are connected with the particle *doncs* 'then'. The king becomes indignant not at the motion but at its aim: he does not want to catch his own relative. This pragmatic function is very similar to that in the following example, which reflects a current Spanish use of the GO-construction: it can express a reaction of surprise to a suggested proposition. Consider (98).

(98)

¿Te acuerdas de tus sueños? – ¿No **me** **voy** *a* **acordar?**
                                Neg Refl Aux.1Sg.Pres Prep remember.Inf
'Do you remember your dreams? – **Why shouldn't I remember?**' (of course I remember)

In Example (98) the sense of futurity is eclipsed and the emotional function of the GO-construction gains ground. Example (98) shows such a use of the modern Spanish *ir a* + Inf in which surprise and indignation are expressed to a proposition which has been suggested before (here: that the hearer does not remember her/his dreams) and which the subject believes to be false. Thus, with the use of this construction in a negative question form, the speaker indirectly expresses her/his surprise and indignation, and, at the same time, denies the suggested proposition. The similarity of Catalan examples like that in (97) and the current Spanish function of the periphrasis provide further evidence in favour of their common origin, because it shows that both Romance GO-periphrases had probably the same functions in their origins. Some of them have survived in Spanish (e.g. intentional modality, emotional charge) and some of them in Catalan (e.g. completedness of the action, past tense value). The development of these functions may have taken place in contexts of different grammatical systems and morphological environments.

The question arises as to what might have been the reason for the divergence in the later semantic development of the Catalan and Spanish constructions? Why did the Catalan construction abandon the common path of semantic change of purposive motion verb constructions? This question leads us back to the problem of which morphological version of the periphrasis should be taken as our starting point in the analysis: the present tense or the past tense version? Notice that it makes a significant difference whether intention is indicated in the present or in the past because different inferences can be drawn in each case. The intention in the present makes a future action expected, while the intention in the past suggests that the performance of the action has been completed. Since human beings are more interested in goals and purposes than in the intention itself, an inference of a future action will be drawn. This can be a real future action or 'future in the past'. The following patterns of inferences summarize this idea:

(99)

Intention in present tense ('I want', we can infer future actions):

intention of doing something → action expected, the intended action is likely to happen in the future → future tense

(100)

Intention in past tense ('I wanted', we can infer past actions):

intention in the past of doing something → action expected in the past, the intended action is likely to happen in a posterior moment in the past → certainty of the past performance of the action → past tense

According to this analysis, both Catalan and Spanish medieval GO-constructions instantiate a process whereby a motion verb grammaticalizes into a marker of intention. At this stage of their evolution they both seem to convey the same meaning. Their later semantic divergence, namely, in Spanish to a future and in Catalan to a past tense, can be attributed to the tense and aspect of the verb 'go' in the most frequent contexts of use, which induced different inferences. On the basis of our common world knowledge we can assume that, on the one hand, actions are usually preceded by intentions, and, on the other, that intentions are usually followed by actions. The inference that the subject intended the future occurrence of the proposed action is the first step in a process whereby GO can evolve into a future or a past tense. The intermediate stage may be that of 'prediction', because 'the attribution of an intention to a third person can, in context, imply a prediction on the part of the speaker' (Bybee et al. 1994: 254). I argue that whether such a construction evolves into a future or a past tense depends on the tense of the GO-verb, which determines the point in time when the predicted action was, or presumably will be, performed.

In the methodological chapter, I intended to show how metalinguistic considerations can aid decision-making between competing hypotheses in linguistic research. After taking into account methodological considerations and the results of the comparison of hypotheses in the literature, I have accepted the approach according to which we should take occurrences of the periphrasis with the verb *anar* in the preterit as the starting point of the analysis. The semantic evolution to a past tense can be attributed to the preterit context in which the grammaticalization of the Catalan construction may have taken place. In a past reference context, the use of the Catalan periphrasis suggested that the action had been completely performed. Because of the frequent co-occurrence, a strong associative link formed between the use of the periphrasis and the perfectivity of the action, and, in the end, this perfectivity was integrated into the coded meaning of this verbal construction. Semantic change may have taken place at this point of the grammaticalization of the construction, while the spread of the present tense auxiliary version seems to be a later phenomenon, linked to a more advanced stage of grammaticalization. Such an interpretation of verbal forms conjugated in the present tense is also suggested by the analysis of occurrences found in our historical corpus.

4.2 Pragmatic inferencing in the grammaticalization of the Catalan *anar* + Inf construction

In Section 2.4.2, when discussing the role pragmatic inferencing (and especially conversational implicatures) play in semantic change in grammaticalization, I made a distinction between two processes. On the one side, pragmatic inferencing can promote the salience of a certain meaning component within the meaning of a linguistic unit under grammaticalization, and, on the other, implicatures can become part of the semantic meaning of a certain form, attached to a certain construction as a new semantic component. I argue that both processes are present in the grammaticalization process presented in this book. The first mechanism may be at work in the salience of the semantic component 'intention' within the meaning of 'go'. Since the meaning component 'intention' is a content that originally forms part of the meaning of the purposive construction formed with GO, it does not have an implicature origin. The same meaning component may be present in other cases where it is not associated with a grammaticalization process. After the meaning component 'intention' had become salient within the 'motion' meaning, similar inferences became possible, as in the use of periphrases which originally mean 'intention'. 'Intention' in the past suggests that the intended action was actually performed, which led to the formation of a past tense. Consequently, we have to make a distinction between the following: where does the new meaning originate from, and why does it become salient? Although 'intention' is a content that originally forms part of the meaning of the purposive construction, the new intentional meaning could become salient and develop due to several pragmatic processes. Thus, we can distinguish between two main stages in the grammaticalization of *anar* 'go' + Inf: (a) the salience of the semantic component 'intention' and (b) the semanticization of an implicature, which refers to a completed action. The content of completedness or perfectivity present in the new meaning may originate from outside, relying on inferencing mechanisms in context, based on world knowledge and information concerning discourse-structuring.

4.3 Further steps in the grammaticalization of the Catalan *anar* + Inf construction and possible typological parallelisms

Finally, although I have only aimed to provide an analysis of the initial stage of the grammaticalization of the Catalan *anar* 'go' + Inf, it would be worth briefly considering later stages of this grammaticalization process, too, in order to have a complete picture. We should try to answer whether its later development also fits any cross-linguistically recurrent tendency of semantic change, or whether it must be understood as a language-specific process. Several authors assume that the Catalan construction was used at a stage of its development with a discourse-structuring function. In Section 2.1.1 I have already discussed possible typological parallels of the evolution of the Catalan GO-construction. In this section I examine whether semantic change processes at a later stage of grammaticalization can be paralleled to processes detected in other languages of the world.

A possible use of the Catalan construction as a discourse-structuring device has been hypothesized several times in the literature. Colon (1959/1978: 125–127) attributes a kind of emphasizing function to the construction: it may indicate the part of the discourse considered to be the turning point of the story. Pérez Saldanya (1996: 84) analyses occurrences of *anar* + Inf at an intermediate stage of grammaticalization as instances of a narrative-aspectual construction, whose function is to segment the discourse (cf. Section 2.3.2). A similar supposition is provided in Detges (2004), who claims that, at a later stage of the evolution of *anar* + Inf, the exaggerated use of the construction in the inchoative meaning led to a loss of efficiency, and due to this iterated use in the same context it became a discourse-structuring device. According to Detges's (2004: 218) hypothesis, the construction conventionalized into a foreground marker, which marked the most important act in a discourse (cf. Section 2.3.3). Finally, Segura-Llopes (2012) also supposes a discourse-structuring function in the medieval use of the construction and associates these occurrences with the use of the construction *ir a* + Inf in current Spanish as a 'demarcative periphrasis'. This use of the Spanish construction demarcates the final event in a series of events, which is represented as unforeseen and surprising (see Bravo Martín 2008: 53–79).[2] Although this semantic nuance can be detected in some medieval occurrences of the Spanish *ir (a)* + Inf, the authors emphasize that it is a construction independent from that from which the immediate future tense developed (Bravo Martín 2008: 53). According to the analysis provided in Segura-Llopes (2012), the Catalan *anar* + Inf construction conveyed a similar demarcative meaning at a certain stage of its evolution and was used to mark pivotal events in the narration.

All these uses belong to a discourse-structuring function and can be associated with the tendency of verbs meaning 'go' or 'come' to grammaticalize into textual connectives. This grammaticalization cline is addressed in Bourdin (2008), who discusses various grammaticalization processes where verbs meaning 'go' or 'come' developed into markers of textual connectivity. Bourdin (2008) collected data from a sample of 64 languages, which allows him to observe that this grammaticalization pathway has a cross-linguistic foundation, although it is disproportionately widespread in African languages. Bourdin distinguishes between two groups: in some languages the directional deictic grammaticalized into a marker encoding simple sequentiality, while in others it fulfils other discourse functions, such as the following: culminativity, foregrounding, or counter-consecutive meaning. It is interesting that if some languages employ 'go' as a specific type of marker, it may be predicted that some other languages employ 'come' to fulfil the same function.

Heine and Kuteva (2002: 68–69, 156–157) also mentions the grammaticalization cline discussed by Bourdin in their *World Lexicon of Grammaticalization*. The authors describe the semantic changes '"come" > CONSECUTIVE' and '"go" > CONSECUTIVE' as instances of a more general process whereby process verbs are grammaticalized into markers used to structure narrative discourse. These linguistic units, before developing this consecutive meaning, are often used to introduce new, unexpected events. Bourdin (2008: 41) notes that the verb 'go' in the Catalan *anar* + Inf construction moved along this grammaticalization cline up until the final stage, i.e. a past tense meaning. Bourdin reconstructs its history following Detges (2004): this verb became a past tense auxiliary and at an intermediate stage in its grammaticalization it functioned as a narrative connective emphasizing the unforeseen nature of the event described in the infinitive. Let us examine whether the development of the Catalan *anar* could really be reconstructed in such a way.

Bourdin (2008) links the evolution of the Catalan *anar* to semantic changes of 'go' and 'come' in some varieties of spoken Arabic. The verb 'go' developed into a counter-consecutive (cf. Example 101) or sequentiality marker in spoken Egyptian Arabic. As mentioned above, sequentiality markers encode the temporal succession of events. In contrast, counter-consecutive markers, beyond sequentiality, highlight the unforeseen nature of the event as well. The meaning of these latter markers is therefore the opposite of that of consecutive markers, which specify the event as a consequence of events previously described. In spoken Lebanese Arabic the verb originally meaning 'go' merely signals that the event is situated in the past (cf. Example 102). This latter use seems to be more similar to the use of *anar* + Inf in current Catalan. Let us consider Bourdin's (2008: 41–42) examples from these language varieties:

(101)

ra(a)ḥ   *Darab-ni.*

COUNTERCONSECUTIVE [= **go**].PRF.3SG.M³ strike.PRF.3SG.M-OBJ.1SG

'He suddenly struck me.'

(Mitchell and al-Hassan 1994: 77, spoken Egyptian Arabic)

(102)

ʕād   reḥ-na   *šef-nɛ-h*   *mbɛreḥ.*

again   **PST** [= **go**].PRF-1PL   see.PRF-1PL-OBJ.3SG.M   yesterday

'We saw him again yesterday.'

(Feghali 1928: 8, spoken Lebanese Arabic)

Occurrences in (101) and (102) are different from a morphological point of view from examples of the Catalan *anar* + Inf: in both the spoken Egyptian Arabic (101) and the spoken Lebanese Arabic (102) examples, the verb 'go' is combined with a verb in the perfect. As for the meaning, however, they can be paralleled with two stages in the grammaticalization of the Catalan *anar* + Inf construction. The occurrence in (101) is worth comparing with preterit occurrences of the Catalan *anar* + Inf with the infinitive *ferir* 'to attack, hit, wound'.

Unfortunately, Bourdin's paper does not offer details of the historical developments of these constructions, thus it is not clear what kind of contexts and inferences enabled the evolution of this kind of meaning, neither it is possible to decide whether these processes belong to the same grammaticalization cline as the development of the Catalan construction. Nevertheless, at first sight they seem to be parallel to some medieval occurrences of Catalan *anar* + Inf.

The case most similar to the semantic change of the Catalan construction seems to be when 'go' or 'come' develop into a discourse connective that encodes a combination of sequentiality and purposiveness (cf. Bourdin 2008: 48–49). The following Akan example (103) quoted in Bourdin (2008: 49) along with purposiveness also encodes sequentiality:

(103)

mé-kɔɔ nkran kɔ-tɔɔ ntamá.

1SG-**go** Accra **go**-buy cloth

'I went to Accra in order to buy cloth and I did.'

(Sebba 1987: 193, Akan)

On the basis of the meaning indicated by Sebba, it seems that the construction also conveys the information that the act of buying cloth was performed. Thus, of all the cases provided in Bourdin (2008), this one seems to be the closest to Catalan, albeit the semantic trajectory followed by the Akan verb is not detailed enough.

In sum, Bourdin (2008) collects cases of grammaticalization of motion verbs into markers of textual connectivity. The textual connectives concerned may encode sequentiality, culminativity, the unforeseen nature of an event, or a counter-consecutive meaning, among others. In some cases, they even grammaticalize into past tense markers. Bourdin mentions the case of Catalan together with some varieties of spoken Arabic, and he also refers to the possibility of the same process in some Bantu languages. Unfortunately, in the appendix listing languages with a sequential derived from 'come' or 'go', we do not find any reference concerning these cases, since Bourdin only deals with this grammaticalization cline until the development of the discourse-structuring function. This is the reason why the Catalan perfective past is not included in the appendix: only the Old Catalan 'go' appears in it, as a marker encoding suddenness (counter-consecutive meaning). Bourdin (2008: 41) relies on Detges (2004) in the characterization of the Catalan construction *anar* + Inf. Therefore, the grammaticalization cline he accepts could perhaps be reconstructed in the following way, ending up with a past tense meaning:

> motion verb > marker of inchoative aspect > textual marker encoding the unforeseen nature of the event > past tense marker.

I have rejected above the inchoative hypothesis provided in Detges (2004). On the basis of the same data, we can also call into question the assumption which is inherently present in Bourdin's analysis, namely, that the perfective past meaning would be the final point of a grammaticalization cline whose intermediate stage corresponds to an inchoative meaning.

Heine and Kuteva (2002: 156–157) mentions the verb 'go' as a possible source of consecutive meaning, but their examples do not reveal whether this semantic change process has an intermediate stage, and, if so, what it would be. They mention examples of this process from Moré and Kxoe languages, where these verbs are grammaticalized into markers used to structure narrative discourse; however, details are not provided. Detges (2004), who assumes a similar discourse-structuring function for the Catalan construction, traces it back to inchoative meaning. Neither Bourdin (2008) nor Heine and Kuteva (2002) mention such a development of inchoative meanings in the world's languages, and it is not assumed in the spoken Arabic varieties presented above, either. Detges (2004: 217) exemplifies the discourse-structuring function of the Catalan construction with a fragment from Muntaner's Chronicle, which is characterized by iterated uses of the periphrasis. However, it is not obvious whether we really are dealing with a discourse-structuring function at that point in the text.

I agree with Bourdin's (2008) statements in the sense that the perfective meaning of the Catalan construction may represent the final stage in a grammaticalization cline where a motion verb comes to be used as a discourse-structuring device at an intermediate stage. A comparison of occurrences of the medieval Catalan *anar* + Inf and the Arabic examples provided in Bourdin (2008) reveal that the historical development of these formally different constructions can be paralleled from a semantic point of view: in spoken Lebanese Arabic (Example 102), Akan (Example 103) and the languages mentioned in Section 2.1.1 – although they are poorly documented – similar to the case of Catalan, a verbal element originally meaning 'go' developed into a marker of perfectivity or past tense (cf. also 'go' completive in the Tucano language in Section 3.6.6). The development of the Catalan construction can be described as fulfilling a discourse-structuring function at a certain stage of its evolution, but it would be problematic to derive it from the inchoative aspect. However, it can be associated with pathways presented in Bourdin (2008), inasmuch as we start from the function assumed by Segura-Llopes (2012) and Colon (1959/1978). According to such an account, the early Catalan construction indicates pivotal events in a narrative. In this sense it can be paralleled with counter-consecutive markers, as, for instance, is exemplified by the spoken Egyptian Arabic occurrence in (101).

Finally, one question remains: which point of the grammaticalization pathway of the Catalan construction does this discourse-structuring function belong to? The study presented in this volume has revealed that the Catalan *anar* + Inf construction may have become an intentional periphrasis at a certain stage of its evolution. At this stage the iterated use of the periphrasis within the same context is very rare, and it is not widely used in narrative texts, thus the discourse-structuring function may be hypothesized to emerge later. In order to reveal exactly how it developed, further research is needed, also relying on later texts. Unfortunately, Bourdin (2008: 45) concludes: 'Indeed, in few languages is it possible to reconstruct with any degree of certainty the actual diachronic route which "come" and "go" have travelled as they evolved into sequentials.' Then he describes two possible scenarios in which deicticity is not directly involved: the iteration scenario, when the motion verb developed into a textual marker, echoes a previous occurrence of the same verb used with its lexical meaning, and the futurity scenario, where the use as a sequential is derived from a previous use as a future marker, although there are several counter-arguments to this latter hypothesis.

In summary, at the moment and on the basis of my historical corpus, it is not possible to analyse the development of the discourse-structuring function.

Nevertheless, frequency may have played a crucial role in this process. The intentional construction may have been used to draw the addressees' attention to certain events in the narrative. As a result of the ongoing grammaticalization, the frequency of use may have increased, and this increase may have implied the loss of this discourse function. When the construction appears several times in the same context, the addressee calls into question whether all these events described in the periphrases are interesting and important ones, that is, the use of the construction loses its effectiveness in attracting the addressees' interest. Since the use as a discourse-structuring device is dependent on frequency, it can be dated after the development of intentional meaning.

Relying on the above considerations, we can assume that the case of the Catalan construction is not a single phenomenon: it can be linked to processes revealed in other languages of the world, even if the exact details and formal aspects of these processes are different. These differences can be well described in a framework such as that provided in Fischer (2010) (see Section 3.6.3.3), which attributes a central role to grammatical constraints on the process of grammaticalization. If we assume that grammaticalization is determined by the grammatical endowment of the language involved, it is not surprising that details of a certain type of grammaticalization process are diverse in typologically quite different languages.

4.4 A summary of the proposed hypothesis on the grammaticalization of the Catalan *anar* + Inf construction

Throughout this book, I have argued that the particular pattern of change of the Catalan *anar* + Inf construction is made up of small steps that in themselves parallel changes detected in other languages around the world (i.e. the formation of intentional periphrasis or a discourse-structuring device from a motion verb construction). However, the GO-verb of this originally purposive periphrasis developed under the influence of the grammar of Catalan (cf. the present–preterit syncretism of 1Pl and 2Pl forms of *anar* 'go' in Old Catalan, the analogical processes and the reanalysis of the construction as a periphrasis with present tense auxiliaries (assumed by Juge (2006), discussed in Section 3.6.4.4), the absence of a literary norm during the period known as the Decadence, the decline of the simple past tense etc.). Now, I will give a summary of the hypothesis developed in this volume on the beginnings of the semantic change of *anar* + Inf, highlighting the main statements step by step.

1. The original version of the construction under study was without any preposition. The prepositional version proliferated in Catalan (and perhaps also in Spanish) in the course of time in order to differentiate those cases where the verb 'go' appears as a main verb.

2. The earliest occurrences may have had the verb *anar* 'go' in the perfective past tense in a preterit context (i.e. conjugated as the context required). The present tense occurrences do not reflect a historical present usage but can rather be interpreted as grammaticalized forms. The present tense version of the Catalan GO-construction in a preterit context is hypothesized to be a later development, which may have proliferated as a consequence of the ongoing grammaticalization of the construction.

3. Therefore, when investigating the beginnings of semantic change, one has to take into account the preterit, non-prepositional occurrences.

4. At an intermediary stage of grammaticalization, the construction became an intentional periphrasis, as in other GO-constructions in various languages around the world; that is, it expressed the intention of the subject and not the inchoative aspect.

5. In some early contexts, although the perfectivity of the action was not explicitly expressed, it had to be inferred in order to have a coherent discourse. The frequent use of the construction in this type of context might have established a strong associative link between the use of the periphrasis and this inferential content. The perfectivity inferable in most contexts of use might have played a decisive role in the semantic development of the Catalan *anar* + Inf construction.

6. The further development of the Catalan construction can be described as fulfilling a discourse-structuring function at a certain stage of its evolution, which can be paralleled with counter-consecutive markers. This function may be hypothesized to emerge later than the intentional meaning; however, it is outside the scope of the present research.

CHAPTER 5

Summary

5.1 Object-scientific findings of the study

5.1.1 Topic and methods of the study

The study presented in this book has aimed to provide a comprehensive analysis of the initial stage of grammaticalization of the Catalan *anar* 'go' + Inf construction, investigating it from a historical pragmatic perspective. The evolution of this medieval Catalan construction is worth investigating, because the result obtained is different from what can be detected in the grammaticalization of similar purposive constructions with motion verbs in various languages around the world. In this book I have shown that it is possible to provide an analysis that describes the beginning of the grammaticalization of the Catalan *anar* + Inf construction as an instance of a cross-linguistically recurrent tendency, inasmuch as it followed the pathway of the semantic change of GO-constructions detected in several languages around the world. And we might say the same about later stages of this process.

The study has been carried out mainly by taking into consideration occurrences of *anar* + Inf found in historical texts from the 13th to the 16th century. The Catalan *anar* 'go' + Inf shows a similarity of contexts, functions, semantic nuances, and morphological characteristics with the Spanish *ir (a)* 'go' + Inf construction in the Middle Ages. This observation has suggested that they have a common origin and has prompted me to compare both constructions. Such a comparison has seemed to be useful, because the evolution of the Spanish GO-construction fits a cross-linguistically recurrent tendency of semantic change of motion verb constructions. The formation of the Spanish immediate future tense from the GO-construction *ir (a)* 'go' + Inf is a typical representative of a well-documented process with various typological parallelisms. In contrast, the development of the Catalan construction has not followed any typical grammaticalization clines; moreover, it lacks well-documented typological parallelisms.

5.1.2 Morphological analysis

Several occurrences found in the texts of my corpus still reflected the original meaning of the Catalan construction *anar* + Inf and Spanish *ir (a)* + Inf: these GO-constructions were originally purposive constructions, which referred to a motion with the aim of carrying out a certain act: 'go in order to do something'. However, other examples already seem to belong to more advanced stages of grammaticalization.

Some morphological characteristics of the medieval constructions are different from the properties of the modern ones. First, in some medieval occurrences the verb 'go' and the infinitive are linked together with the preposition *a* 'to'. Second, the verb 'go' can appear in the medieval Spanish and Catalan texts conjugated in either the past or the present tense. As a first step, I have investigated the distribution of these morphological versions.

The study of the use of the preposition *a* 'to' in historical texts has provided the following results. Occurrences in which the purposive preposition *a* 'to' appears between the finite verb 'go' and the infinitive are very rare at the beginning, but later their use is extended. The prepositional version of the construction seems to have a more restricted use in the medieval language state than the use of the construction without a preposition, and it tends to appear in later texts. It may be hypothesized that the original version of the construction under study was without any preposition. Therefore, when investigating the semantic change, we have to take into account the non-prepositional occurrences. On the basis of the results of my study and those of other authors, it may be supposed that the prepositional version may have proliferated in Catalan in the course of time in order to differentiate those cases where the verb *anar* appears as a main verb, with the meaning 'go'. A similar process is assumed in Radatz (2003) concerning the construction in Spanish where the prepositional variant (*ir a* + Inf) grammaticalizes into an immediate future tense. If we also take into consideration the other supposition of Radatz (2003), according to which also the Catalan *anar a* + Inf construction is currently developing an immediate future meaning, then we should suppose the same processes in both languages, although with a discrepancy in time. Therefore, the spread of the prepositional form reflects the ongoing grammaticalization of the construction without a preposition.

As regards the verbal tense in which the verb 'go' is conjugated, the following results have been obtained. Differences in the evolution of the Catalan and the Spanish constructions could already be detected in the medieval language state, as manifested in the distribution of the present and perfective past tense versions of the Catalan and the Spanish GO-constructions. The analysis of occurrences in their contexts has revealed that the present tense version of the Catalan GO-construction in a preterit context is a later development, which may have spread in consequence of the ongoing grammaticalization of the construction, and the verb *anar* 'go' may have appeared in the perfective past tense in the earliest occurrences (see Juge 2006 and 2008). It can be argued that the present tense occurrences do not reflect a historical present usage but can be interpreted as grammaticalized forms. The Spanish texts of the corpus have not reflected the same propagation of present tense occurrences.

5.1.3 Comparison with functionally similar constructions

After presenting the morphological characteristics of the Catalan *anar* + Inf construction, I have examined the path of semantic change it followed, taking as a starting point the preterit variant of the construction without a preposition. As a result of this research, a somewhat different picture has emerged of the semantic change in the grammaticalization of *anar* + Inf from that presented in the previous literature.

Most of the relevant authors have derived the current Catalan *anar* + Inf construction from a historical present usage (Colon 1959/1978, 1976/1978; Badia i Margarit 1981; Pérez Saldanya 1996; Pérez Saldanya and Hualde 2003; Segura-Llopes 2012), while others have argued that the present auxiliary occurrences do not show the characteristics of a historical present use and do not retain the lexical meaning 'go' of *anar* (Bruguera 1981; Juge 2002, 2006, 2008). A further account has hypothesized that the Catalan construction under study was an inchoative periphrasis in the medieval language state (cf. Detges 2004). Finally, Calvo Pérez (1995) and Martos (1996) have analysed the semantic change of *anar* + Inf starting from the deictic nature of the verb *anar* 'go'.

Taking into account the morphological findings presented above, the results discussed in relevant literature and the similarity in uses of the Catalan and Spanish GO-constructions in the language stage under study, I have analysed historical occurrences in the following way. I have based my analysis on occurrences without a preposition, with the finite verb conjugated in the perfective past. I have started from the assumption that the initial stage of the semantic change of the Catalan construction followed the typical grammaticalization cline from motion verbs to future markers observed in various languages in the world, where the meaning 'intention' is the intermediary stage of the process (cf. Bybee and Pagliuca 1987; Bybee et al. 1994; Bybee 2002: 181; Heine and Kuteva 2002: 161–163). In order to test this hypothesis, in addition to the Catalan and the Spanish GO-constructions, I have included medieval Catalan and Spanish occurrences of the periphrasis *pensar (de)* 'think of' + Inf in the analysis, which expresses the subject's intention to carry out the action described in the infinitive. I have also considered the hypothesis provided in the relevant literature, according to which the Catalan construction under study was an inchoative periphrasis in the medieval language state. In order to test this hypothesis, I have included occurrences of the inchoative construction *començar (a/de)* 'begin to' + Inf in medieval

Catalan and Spanish. Although the THINK OF-periphrasis has an intentional meaning and the BEGIN TO-periphrasis has an inchoative meaning, both seem to be used in similar contexts. Therefore, the integration of these pieces of data has seemed to be justified by the similarity of the contexts and functions these periphrases seem to have in medieval texts (cf. Fischer 2007: 15). The corpus of the study thus contained a total of 1,903 occurrences, namely, 814 occurrences of the medieval Spanish and Catalan GO + Inf periphrases, 221 examples of the THINK OF + Inf periphrases and 868 of the BEGIN TO + Inf constructions.

I have presented the types of infinitive that appear in these three constructions in the medieval language state by analysing entrenched local schemas, because I have supposed that the semantic change of the Catalan motion verb construction might have emerged from these very frequent uses. I have examined the use of infinitives in these constructions relying on a thorough contextual analysis, in order to detect similarities and differences. The comparison of the uses of these three periphrases has revealed a certain type of functional similarity: all three constructions have been used in order to attract the listeners' attention. However, a more detailed qualitative analysis of the contexts has made obvious a closer similarity between the uses of *ir (a)/anar* 'go' + Inf and *pensar (de)* 'think of' + Inf, and a difference of both from the use of *començar (a/de)* 'begin to' + Inf. The verbs *començar* 'begin', *pensar* 'think', and *ir/anar* 'go' have all occurred with similar types of infinitives, but they have added different nuances to the description of events. In the case of the BEGIN TO-construction the event in the infinitive has been described as a durative one, but which has been interrupted as soon as it has begun. With this strategy the subsequent wedged-in events have been emphasized. In contrast, in the case of the GO- and THINK OF-periphrases the event described in the infinitive has been highlighted, which has been referred to as a whole, completed, perfective event with considerable consequences.

These findings throw a different light on the history of the Catalan GO-construction under study and relate its origins to the salience of the meaning component 'intention', suggesting that Detges's (2004) inchoative aspect hypothesis concerning the origin of the Catalan GO-past is mistaken. The historical analysis rather suggests that occurrences of the Catalan *anar* 'go' + Inf refer to completed actions and not to their beginning. GO + Inf in medieval texts describes the action as a whole, and its occurrences are often followed by a description of consequences. The perfectivity inferable in most contexts of use might have played a decisive role in the semantic development of the Catalan *anar* + Inf construction.

To sum up, I have argued that, despite the different endpoints of development of the two Romance GO-constructions under study, their histories run parallel for a time. Both the Catalan *anar* 'go' + Inf and the Spanish *ir (a)* 'go (to)' + Inf had a stage in their history when the meaning component 'intention' was salient within the semantic structure of the verb 'go'. At this stage, the construction expressed the intention of the subject and not the inchoative aspect, as supposed by Detges (2004). The salience of the meaning 'intention' guaranteed the dynamism in the description. The historical analysis suggests that the originally purposive construction *anar* 'go' + Inf at the beginning of its grammaticalization did not express the inchoative aspect, but rather became an intentional periphrasis, as other GO-constructions in various languages around the world.

5.1.4 Contextual analysis

Regarding the topic of the text, descriptions of battle scenes contain most of the utterances with these GO-periphrases, where they seemed to have fulfilled a function of dynamization, intensifying and drawing the addressees' attention to the event which speakers consider to be the most important one. This is the most typical context. However, taking further criteria into account, various types of context could be differentiated, which, as they can be linked to successive stages of grammaticalization, helped me to reveal the process of semantic change. When attributing meaning, I have followed the methodological principle, according to which, if it is possible, i.e. if nothing in the context contradicts it, we should attribute the original lexical meaning of *anar*. Quantitative data have suggested an ongoing increase of occurrences over time where the lexical meaning can be excluded.

When revealing the process of semantic change, I have started from the occurrences which could be interpreted literally. Within the group of contexts where the verb *anar* occurred in its source meaning, I have differentiated between two main types. In the first type, the motion meaning could be assigned exclusively, while in the second type a semantic ambiguity could be revealed. Semantic ambiguity is one of the most important characteristics of contexts in semantic change, because it triggers pragmatic inference mechanisms. Thus, I have concentrated on these semantically ambiguous contexts, where both the original meaning and a new meaning could be assigned.

Some of the earliest occurrences already suggest that the construction might have possessed another function beyond conveying motion in space. It is also important to mention contexts where, although the literal interpretation of the verb *anar* has been possible, expressing motion in space seems to be redundant. In other contexts, the discourse would have remained coherent and meaningful with the omission of the verb *anar*, i.e. the addressee has been allowed to interpret the verbal construction in another way. These types of contexts may have triggered semantic change.

However, grammaticalization may have been supported by other factors, too. Occurrences with a possible literal meaning could be grouped in another way. In some contexts, the utterance containing the occurrence of the GO-construction continued in a way that allowed the addressee to conclude that the goal of the motion expressed in the infinitive was thoroughly performed, while in other contexts, this conclusion was not possible. In the former contexts, although the perfectivity of the action was not explicitly expressed, it had to be inferred in order to have a coherent discourse. The frequent use of the construction in this type of context might have established a strong associative link between the use of the periphrasis and this inferential content. Finally, another type of context should be mentioned: where the motion described by the verb *anar* was presumably performed in a very narrow space. Moreover, some contexts belonging to this latter type have given rise to the question of whether it is reasonable to suppose the literal meaning at all. In those cases, the discourse would have remained meaningful without supposition of any motion. These contexts differ in this sense from those where the motion covered long distances and led to a different scene.

Analysis of the distribution of these context types over time has led to the following findings. The number of contexts where the 'motion' meaning is possible decreases over the period examined, while the number of contexts where it is excluded increases. This tendency could also be observed if we interpret the present tense form as instances of the historical present, although it is more pronounced when rejecting the historical present hypothesis.

I have examined the correlations between context types and verbal tense in a similar way. In the case of preterit tense versions, I have identified a decrease in the use of context types where it is impossible to assign the 'motion' meaning. I suppose that this is because its function has been taken over by the present tense version. Present tense versions can be associated with context types which exclude the original, lexical meaning of *anar*.

5.1.5 Deciding between rival hypotheses

At this point of the research, relying on my own results and the findings presented in relevant literature, I had to decide between competing hypotheses concerning the tense of the finite verb of the periphrasis: whether we have to take the present or the past tense GO-verb as the starting point in the description of semantic change. I have relied on methodological considerations and used the so-called p-model as presented in Kertész and Rákosi (2012). The findings of this part of the research have made plausible the hypothesis according to which the occurrences with a past tense GO-verb should be the starting point in the description of the grammaticalization process. In contrast, the historical present hypothesis, which emphasizes the role of present tense versions, has turned out to be implausible.

Taking everything into consideration, I have characterized the initial stage of the evolution of the Catalan *anar* + Inf construction in the following way. At the initial stage of their evolution, both the Catalan and the Spanish medieval

GO-constructions instantiated the grammaticalization of a motion verb into a marker of intention. Their later semantic divergence, namely, in Spanish to an immediate future and in Catalan to a past tense, can be attributed to the tense and aspect of the verb 'go' in their most frequent contexts of use, which induced different inferences (for more details, see Nagy C. 2010a; cf. also Pérez Saldanya 1996). The preterit context in which the grammaticalization of the Catalan construction may have taken place was different from that in which future tense markers usually evolve. In a past reference context, the use of the Catalan periphrasis suggested that the action had been completely performed and that the performance of the action was more dynamic and intentional than usually the case. Because of the frequent co-occurrence, the pragmatic function (of structuring the discourse by evoking dynamism and attracting the addressees' attention to central events in this way) was lost, while a strong associative link was formed between the use of the periphrasis and the perfectivity of the action, and, in the end, this perfectivity was integrated into the coded meaning of this verbal construction.

As for the typical pathways of semantic change, on the basis of the above considerations, it may be hypothesized that the evolution of the Catalan GO-construction reflects a cross-linguistically attested tendency, inasmuch as it moved along the following cline until the 'intention' stage: motion with the aim of doing something > intention > (future). The further evolution of the Catalan construction can also be related to a cross-linguistically recurrent tendency: the grammaticalization of some motion verbs into sequential markers and, more generally, into textual connectives (cf. Bourdin 2008). The idiosyncrasy of the evolution of the Catalan GO-construction can be attributed to constraints imposed by the grammar of Catalan language.

5.2 Methodological findings of the study

5.2.1 Data types in historical pragmatics research and the notion of historical data

Relevant chapters of this book have discussed the topic of linguistic data and methods in historical pragmatics research. On the one hand, historical linguists often have to cope with the scarcity of historical sources, the question of indirectness and unreliability of data, and the non-accessibility of several data source types. Historical linguistics lacks data from discourse completion tests and experiments, has only indirect access to native speakers' linguistic judgments, and written sources offer only approximate evidence. On the other hand, pragmatics research aims to investigate procedures of production and interpretation that cannot be formalized entirely. In addition, it focuses on levels of meaning which are difficult to access, concerning which data collection is very difficult.

In any case, it is of primary importance that such studies combine various data sources and never stop broadening their database.

During the research, I have combined different data sources and tried to incorporate as many types of data as possible. In the methodological chapter, I discussed difficulties concerning the use of corpus data, which can originate either from using the linguist's intuition or from properties of the historical documents themselves. Although I have primarily relied on data constructed on the basis of a written historical corpus, several other data types have been present in the context of my research (e.g. data gained from the contemporary language system, typological data, data based on the integration of pieces of encyclopedic and metalinguistic information, etc.). In order to supplement my data sources, I have also considered data regarding functionally and formally similar linguistic units that were in use in the same language state, as Fischer (2004, 2007) suggested.

I have also tried to answer the question of what kind of data historical corpus data really are and how they could be integrated into the research process. I have addressed problems concerning the traditional notion of *historical data*, used as a synonym of *occurrences of a linguistic unit in historical documents*. The p-model, as presented in Kertész and Rákosi (2012), has served as a basis for investigating the difficulties concerning corpus-based data. These difficulties can have their origin both in the characteristics of the historical sources themselves and in the use of linguistic intuition during interpretation. The p-model offers a novel notion of *data* with which relevant methodological problems can be treated. According to the p-model, data are plausible statements originating from some direct source. Data have a double structure: they combine an informational content and a plausibility value, which is assigned on the basis of some direct source. I have presented Kertész and Rákosi's (2012) methodological findings concerning corpus data, complementing them with my own observations and adapting them to historical corpus research. I have placed special emphasis on the question of positive and negative evidence, the phenomenon of using a 'substitute competence', and the problem of the indirectness of data in the historical research.

On the basis of all these considerations, it has emerged that neither a corpus nor, naturally, a historical corpus can be considered a fully reliable data source which serves as a source of certainly true statements. In the case of corpus data, we have to refer in all cases to the linguistic intuition of the researcher as an additional data source, and during the research the plausibility value of corpus data is influenced by that of data from other sources, too. Consequently, corpus data should be considered as data originating from an integrated data source. In addition to the historical corpus as primary source, further data sources also play a role, including, among others, the linguistic intuition of the researcher, previous studies on the same topic, and the theoretical framework in which we are working. Thus, the correct term would be *corpus-based data* or *data constructed on the basis of a corpus*.

5.2.2 Contextual analysis and the notion of context in historical pragmatics research

In the methodological part of the book I also discussed another pivot of historical pragmatics methodology, namely, contextual analysis. Semantic change can be conceived as a change of contexts in which a certain linguistic construction can be used, so the description of grammaticalization is intimately linked to the description of the contexts in which it takes place. The nature of the contexts in which particular stages of semantic change processes can be detected has been the topic of various papers on grammaticalization (Diewald 2002; Heine 2002). I have presented Diewald's (2002) and Heine's (2002) context typologies in detail and made a distinction between two main types of context. The first group includes Heine's bridging context and Diewald's untypical and critical contexts. These context types trigger grammaticalization, because the old meaning is only one of the possible interpretations in them and a new meaning can also be assigned. Ambiguity, which characterizes these contexts, is always a necessary stage of meaning change. The other type includes Heine's switch and conventionalization contexts and from Diewald's isolating contexts those which favour the new reading and exclude the old, lexical meaning. This latter group is more important from a methodological point of view, because the occurrence of the lexical item in one of these contexts can serve as evidence for a previous meaning change. In other words, it presupposes a previous reorganization in the semantic structure of the grammaticalizing lexical item.

Contextual analysis in historical pragmatics should be used to gain not only quantitative, but qualitative data as well. This is possible by considering the broad context of historical occurrences, apart from investigating their distributional properties and combinational possibilities with other lexical units. The analysis of the broad context of the occurrences has provided valuable data about the semantic nuances expressed by a linguistic unit, and, as a consequence, about the ongoing process of grammaticalization. So, I have made a distinction between narrow and broad contexts.

In order to avoid interpretations based on our subjective linguistic intuitions and our knowledge of the current language state, we should find some linguistic clues which may help us to reveal what shades of meaning historical occurrences express. Context may contain several linguistic clues which direct us in guessing meanings in historical documents and help us to minimize subjectivity. Linguistic clues which serve as points of orientation in assigning meanings may already occur in the narrow context. In the case of some occurrences of the construction, the infinitive itself and the nouns or adverbs which appear with it, the grammatical person of the verb or the semantic properties of the subject have become semantically irreconcilable with the literal meaning of the finite verb. In other cases, the context has been broadened in order to find the most plausible reading of a historical occurrence. In the broad context, the content of the preceding or subsequent clauses and, in some cases grammatical characteristics (e.g. verbal tense or verbal mood) could also help us. In other cases, an even broader section of the discourse has been taken into account, or some pieces of encyclopedic information have been activated and influenced the interpretation of the construction, excluding the literal meaning.

5.2.3 Argumentation in diachronic research

In the last part of the book, I discussed certain characteristics of historical pragmatics argumentation. Using the metalinguistic framework presented in Kertész and Rákosi (2012), I examined some hypotheses present in the context of my research from a methodological point of view in order to decide between them. First, I discussed the notions and phenomena of frequency and analogy, which play a central role in historical linguistics argumentation. Since they are widely accepted methods and explanations in this field, statements from an argumentation which relies on them can be assigned a high plausibility value.

I discussed three hypotheses concerning the morphological aspects of the history of the Catalan *anar* 'go' + Inf construction, namely, Colon's (1959/1978, 1976/1978), Detges's (2004), and Juge's (2006) accounts. I intended to show how metalinguistic considerations can aid decision-making between competing hypotheses in linguistic research. I would like to highlight once again that I have not presented these approaches in their entirety. I have only taken into consideration claims that concern the medieval present–past alternation in the use of the construction. Even if some imperfections in their argumentations could be detected, these authors all provide very valuable findings on different aspects of the history of this Catalan GO-periphrasis.

There have been hypotheses in the starting context of the research which have supposed a historical present use in those cases where the finite verb of the *anar* + Inf construction was conjugated morphologically in the present tense. Other authors have supposed the opposite; thus, the starting context of the research has been inconsistent. The extension of the context with new pieces of information concerning the historical present use and the modification of the plausibility values of these hypotheses relying on methodological considerations have helped us to eliminate the inconsistency. The claim according to which we are dealing with a historical present use in relevant historical texts has turned out to be implausible, while the negation of this claim has been given a higher plausibility value.

I have tried to show that dealing with methodological questions is not simply a theoretical issue but can be useful in linguistic research practice. By the reconstruction of argumentation processes from a metalinguistic point of view, we have been able to obtain information which has helped us to decide between competing hypotheses.

5.3 Most important new findings of the research

Finally, I will summarize the most important new findings of the study presented in this volume.

1. The present study, investigating the subject from a historical pragmatic perspective, throws a different light on the initial stage of grammaticalization of the Catalan *anar* 'go' + Inf. Findings on pragmatic aspects of the process have offered new information concerning not only the semantic change of this medieval Catalan GO-periphrasis, but also the cross-linguistically recurrent tendencies of semantic change of motion verbs in different constructions.

2. Concerning the methodology of the present study, in addition to the GO-constructions, I have also included in the analysis medieval Catalan and Spanish occurrences of the periphrases *començar (a/de)* 'begin to' + Inf and *pensar (de)* 'think of' + Inf, because of the similarity of their contexts and functions in medieval texts. Relying on data gained in this way, I have related the initial stage of the semantic change of the Catalan construction to the typical grammaticalization cline from motion verbs to future markers observed in various languages around the world, where the 'intention' meaning is the intermediary stage of the process. The fact that the development followed by the Catalan construction fits a cross-linguistically frequent tendency according to the account provided raises the plausibility value of the present hypothesis.

3. I have provided a methodological analysis of the research process presented in the first part of the book, which has influenced the further development of the research, throwing a different light on the development under study. At the beginning of the research, the historical present hypothesis seemed to deserve a high plausibility value, as it is widely accepted in the relevant literature. After taking into account methodological findings, Juge's (2006) hypothesis has become the more plausible, and it has turned out that an account which starts from the occurrences with the verb *anar* in the preterit tense can account for a wider range of occurrences. The final hypothesis has been developed relying on both the object-scientific and the methodological findings of the study. Thus, the utility of taking into account metalinguistic considerations when we decide between rival hypotheses in linguistic research and judge the reliability of certain claims has also been revealed.

Appendix: Occurrences of the Catalan *anar* + Inf and *anar a* + Inf in the corpus

The Appendix lists the occurrences of the Catalan *anar* + Inf and *anar a* + Inf constructions found in the corpus, arranged in chronological order of the texts. It is organized according to the verbal tense of 'go'.

(Jau)

Anar + Inf

(Jau 16) E vengren alí, foch encès en falles, al fenèvol e **anaren escometre** Don Pelegrí e Don Guillem de Poyo, qui tenien la vetla.

(Jau 60) E, quan fo prop d'éls, demanà los nostres; e, quan éls foren venguts a él, dix: 'Firam en éls, que no són re!' E el primer que anch los **anà ferir** fo él.

(Jau 64) E dix: 'El comte d'Ampúries e·ls del Temple **anaren ferir** als de les tendes, e En Guillem de Muntcada e En Ramon **anaren ferir** a la mà esquerra'.

(Jau 64) E **anaren-me pendre** a la regna él e Don Pedro Romar e Ruy Xemenis de Luèsia e deÿen:

(Jau 85) E, cant entraren los cavalers ab los cavals garnits, **anaren-los ferir**. E era tanta la multitut de la gent dels sarraïns, que·ls pararen les lances, e els cavals dreçaren-se per ço car no podien passar per la espessea de les lançes, sí que agren a fer la volta.

(Jau 85) E en tant feyta la volta, tiraren-se atràs .I. poch e **anaren entrar** los cavals tant, que n'hi ach bé de ·XL· tro a ·L·.

(Jau 86) E els altres sarraïns, quan viren que aquel loch havien esvaït los cavalers ab cavals armats e·ls hòmens de peu, **anaren-se amagar** per les cases, cascú con mils podia; e no s'amagaren tant bé que ·XX· mília no n'i morissen a l'entrar,

(Jau 201) E, quan vench al matí, hoïdes les misses, **anaren-se armar** los escuders e gran partida dels cavalers;

(Jau 224) E **anà'ns abraçar** e dix-nos que Déus nos donàs bona ventura.

(Jau 232) E **anaren-se acordar** e a cap d'una hora tornaren a nós e dixeren a Don Blasco d'Alagó que dixés él ço que havien acordat

(Jau 257) E quan vench altre dia, ans d'alba, menys de sabuda de nós, los almugàvers e·ls servents **anaren pendre** Ruçafa, que és a ·II· trets de balesta prop la vila de València.

(Jau 311) Don Ferrando ab los de Calatrava, e ab don Pero Cornell, e don Artal d'Alagó, e Don Rodrigo Liçana **anaren assetjar** Villena

(Jau 274) E nós levam-nos per ell, en nostres cases ben guarnides e bé adobades, a l'entrar que féu, quan fo prop de nós; e anch no·ns volch besar la mà, mas que s'humilià a nós e **anà'ns abraçar**.

(Jau 391) E **anaren-se acordar** e no volgren venir denant nós aquel dia ni en l'altre tro al vespre, que vench a nós Don Bernat Guillem d'Entença.

(Jau 485) e levà's ·I·ª mànega e **anà ferir** en la mar, e de blancha que era tornà tota negra.

(Jau 520) E **anà's gitar** als nostres peus e besà'ls-nos e pregà'ns per Déu que li perdonàssem.

(Jau 525) E l'apostoli era en sa cambra e, quan li dixeren que nós veníem, tot revestit exí; e veem-lo passar denant nós, e **anà s'asseer** en sa cadira.

(Jau) e, al devalar que nós fayem, Barthomeu Esquerdo, qui era adalil, sobre paraules que havia ab ·I· home, **anà'l ferir** del coltel denant nós e entràsse'n en la tenda que nós havíem prestada a Don G. Romeu

(Jau 21) E, sobre açò, **anam assetiar** Muntcada.

(Jau 22) E puys, quan vench al maytí, **anam hoir** la missa a la esglea major d'Alagó.

(Jau 34) E nós aculim-la bé e, quant vench que ach estat ·II· dies, **anam-la veer**.

(Jau 37) E dixem-nós: 'Con la vila tenen éls?' E sempre lexam los cavals als escuders e avalam; e prenguem nostres armes e **anam-los combatre** e tolguem-los la vila.

(Jau 85) E ab aytant los cavalers foren ja bé ·XL· tro a ·L· ab los cavals armats e endreçaren-se contra·ls sarraïns e cridaren tots a una vou: 'Ajuda-nos sancta Maria, Mare de nostre Seyor!' E cridàvem: 'Vergonya, cavalers!' E **anam-los ferir** e esveÿm-los.

(Jau 155) E altre dia bon matí **anam albergar** a Torres Torres

(Jau 156) e tots ensemps **anam assetiar** Boriana.

(Jau 168) E, quan fo passada ora de vespres, **anam-nos deportar** fora la ost;

(Jau 188) e **anam passar** a Albalat e estiguem aquí ·III· dies.

(Jau 219) e entram-nos-en al Pug e **anam albergar** a les Alcubles.

(Jau 226) E Don Pero Corneyl, qui hoí cridar a armes, va pendre ses armes e brocà aenant. E nós cuytam-nos e **anam-lo pendre** per la regna e dixem-li:

(Jau 237) Ab tant **anam-nos gitar** e no volguem descobrir les paraules a nuyl hom qui fos ab nós.

(Jau 252) E puys **anam-nos veer** ab aquels de Nubles;

(Jau 254) e la regina tench la meytat de la Quaresma en Almenara tro a la Pasqua, e nós **anam tener** la Pasca ab ella;

(Jau 320) Ab tant, **anam assetjar** lo castell en la bega

(Jau 323) E nós brocam aprés ell e, al entrar que él feÿa en la tenda, **anam-lo pendre** als cabeyls e traguem-lo'n

(Jau 339) e ab los altres rich hòmens e **anam assetiar** Xàtiva

(Jau 340) E **anam cridar** per la ost que tot hom que hauria paraules ab los moros, meyns que a nós no u demanàs, que fos pres e que l'aduyxessen a nós.

(Jau 343) E nós lexam en la ost ·CC· cavallers e·ls hòmens de peu que y eren e **anam-nos veer** ab ell.

(Jau 434) E en l'altra dia de Ninou, entrada de janer, nós **anam assetjar** Múrcia. E, a l'anar que nós faem ab nostra ost a Múrcia, fom dels primers qui y foren,

(Jau 462) e tirà la primera pedra lo maestre del fenèvol e errà la brigola; e nós **anam pendre** lo fenèvol e tiram e donam tal en aquela brigola, que la caixa li obrim;

(Jau 496) E un dia, mentre que nós estàvem aquí, envià-nos a dir Don Alfonso de Molina que no era ben sa, e nós **anam-lo veer**. E, al tornar que nós feýem de l'Espital de Burgos,

(Jau 540) E, quant vench altre dia matí, nós **anam-lo veer**

(Jau 550) E, quan fom partits de Perpinyà, **anam reebre** la host de Barcelona, ·I.ª partida que·n venia per terra, e trobam-la a la Bisbal.

(Jau 218) E aquels qui eren en la reraguarda dels sarraïns, qui eren dessús los altres, començaren de fugir primers que aquels qui eren denant; e **van ferir** en la devantera los nostres als sarraïns e obriren-los.

(Jau 226) E Don Pero Corneyl, qui hoí cridar a armes, **va pendre** ses armes e brocà aenant. E nós cuytam-nos e anam-lo pendre per la regna e dixem-li:

(Desc)

Anar + Inf

(Desc II 8, 2) **anà assetgar** Leyda

(Desc II 35, 29) E aytantost aquels exiren del aguait ab lurs senyeres esteses e **anaren ferir** en la reregarda de la host dels sarraÿns

(Desc II 43, 28) Sobr·assó vengren-li bé ·X· cavalers justats e **anaren-lo ferir**, sí que·l abateren a terra e aquí murí. Quant viuren que·l rey era mort, pensaren-se'n d'anar per cames de cavals.

(Desc II 50, 17) E·l chomte, quant hac reposat ·I· jorn, l'endemà **anà parlar** ab l'emperador

(Desc II 51, 19) Ab tant lo chomte **anà parlar** ab la emperadriu

(Desc II 55, 25) mès-se la lansa denant, e **anà ferir** lo cavaler alamayn de tal vertut que la lança li passà de l'altra part per mig lo cors

(Desc II 59, 28) e cavalcà e **anà-la veer** a sson hostal

(Desc II 66, 4) E·l rey En Jacme ajustà ses osts e **anà assetgar** aquel castel de Peníscola.

(Desc II 86, 1–2; 3) E puis partí-sse lo parlament e **anaren menyar**. E quant venc a la nuyt, so fo a al vespra de Nadal, el rey **anà vetlar**

(Desc II 87, 4) E puys **anaren-se deportar**

(Desc II 98, 26–27) A aquest conseyl s'acordaren tots e **anaren ferir** e·ls sarraÿns, cavalers e sirvents.

(Desc II 102, 8–9, 9–10) Aquest conseyl han tuit tengut per bo, e **anaren hoir** les misses e confessaren-se. E puis **anaren-se dinar** molt tost

(Desc II 103, 15) quant foren e·l trescol, encontraren-se ab la host dels sarraÿns e **anaren ferir** en éls

(Desc II 104, 4) e **anà ferir** entre·ls sarraÿns

(Desc II 104, 16) e **anaren ferir** en éls

(Desc II 105, 17) e **anà ferir** entre·ls sarraÿns ab tota sa cavaleria

(Desc II 120, 9–10) Quant lo rey serraý se viu así apoderat, **fóu parlar** al rey molts pleyts per tal que·l se pogués levar de sobre.

(Desc II 124, 29) Quant venc lo dimenge matí, **anaren oyr** les misses

(Desc II 125, 19) Ab tant lo vespre fo vengut e **anaren-se reposar**, que ben ho avien obs, que molt avien trebalat aquel jorn.

(Desc II 133, 2) Quant venc l'endemà, que fo lo jorn de sent Steve, **anaren cavar** al mur

(Desc II 134, 11–12; 12) E·l rey d'Aragó **fóu saber** a tuit per la host que·s aparelassen d'entrar en la ciutat. E tuit **anaren oyr** les misses, e confesaren e conbregaren; a puis **anaren-se dinar** de plors, e de làgrimes e d'altres menjars, e perdonaren-se los uns als altres.

(Desc II 134, 19–20) Ab aytant ·I· sirvent qui era de Barcelona **anà desrengar** ab ·I· penó que portava

(Desc II 136, 13) E quant venc l'endemà matí, que foren levats, **anaren oyr** les misses e puys tornaren dinar;

(Desc II 138, 11) (de la qual hac ·I· fil qui hac nom N'Anfós, al qual fóu jurar tota ça terra quant **anà conquere** la ciutat de Malorches)

(Desc II 145, 26) E mantinent aparelà-sse ab molt gran host e **anà assetjar** la ciutat de València.

(Desc II 148, 8) Puys **anà asetgar** Exàtiva

(Desc II 167, 11–12) e·l comte Jordà, qui avia la devantera, **anà ferir** en la primera escala dels picarts

(Desc II 167, 23) e mès-se l'escut denant e la lança, e **anà ferir** en la pressa dels francescs

(Desc II 178, 8) Adoncs **anaren-se aparelar** tot privadament

(Desc II 181, 14) E lo serv a qui el rey avia perdonat, plen d'ira a de mal talent, **anà pendre** aquel serv del rey a la gola e vulia'l estrangolar

(Desc III 6, 20) s'aparelà ab moltes gens a cavayl e a peu, e **anà assetjar** Múrcia

(Desc III 23, 8) e ajustà gran host de cavalers e d'òmens a peu e **anà assetgar** ·I· castel molt gran qui ha nom Antiló,

(Desc III 26, 14) sí que del tot la refusà e **anà's amagar** en ·I· camp de blat.

(Desc III 30, 15) E puys **anaren assetjar** ·I· casteyl qui à nom Montbauló,

(Desc III 38, 20) lexà tota la altra terra que·ls sarraÿns tenien e **anà asetyar** ab grans hosts, que ac fetes justar, la vila de Muntesa.

(Desc III 39, 29) E quant venc lo matí, tota la host fo aparelada, e les dues parts de la host **anà combatre** la vila

(Desc III 48, 25) Mas N'Esquiu de Miralpex quant fo desots lo pont hon l'ayga corria pus fort, lo cavayl li devec, e **anà's pendre** sus alt a la pila del pont e tenc-la molt fort abrassada, tot garnit que era.

(Desc III 73, 1–2) Sí alsà la ·I· la massa que portava e **anà-li'n donar** ·I· gran colp per les espatles.

(Desc III 93, 24; 26) E quant fo defora, puyí lo cavayl dels esperons e **anà ferir** entre·ls sarraÿns, sí que·l primer colp n'abaté, ab lo pits del caval seu, ·IIII· a terra; e **anà ferir** ·I· sarraý de la lança, que la dàraqua li passà e tot lo cors, sí que l'abaté mort a terra. E puys arancà-li la lança del cors e, de manén, va'n donar tal colp per mig lo pits a aquel qui portava la seyera, que mantinent l'abaté mort a terra.

(Desc III 100, 21–22) e els menjars foren aparelats, e menjaren alegrament e puys **anaren-se reposar**, que la mar los avia trebalats.

(Desc III 104, 6–7) Ab tant les taules foren meses e el rey **anà menyar** ab sos cavalers;

(Desc III 109, 1–2) Quant aquels del mur o agren entès, **anaren-ho dir** al capitani, ser Alaymo;

(Desc III 124, 27) E cascuns **anaren-se reposar e solassar** là hon li plac;

(Desc III 143, 6) E l'almirayl **anà parlà** ab lo rey e comtà-li tot so que lur era esdevengut;

(Desc III 144, 10) sí que·ls almugàvers no·s pogren gandir d'éls, e axí **anaren-se mesclar** ab eyls ensems.

(Desc III 146, 22; 24) venc en ·I·ª plassa, e aquí trobaren ·I·ª gran compaya de cavalers francès e proensals e **anaren ferir** en eyls, sí que·n enderrocaren molts de lurs cavayls e·n alcieren. E En Bernat de Peratalada **anà ferir** ·II· cavalers francescs a través ab la lança, sí que amdós los passà e·ls abaté en terra morts.

(Desc III 146, 13) Mas los almugàvers **anaren trencar** les lisses e el mur de la vila, e entraren dins e los cavalers ab eyls, e meteren foc a la ·I·ª part de la vila;

(Desc III 147, 4) Ab aquestes paraules En Pere Arnau de Botonac **l'anà abrassar** per mig dels flancs, e trasc-lo de la sela tot armat

(Desc III 150, 7) – Certes – so dix lo príncep –, so·m plau. Ara, donques, anats-vos armar, e veuren aquest què sabrà fer. Ab tant lo cavaler **s'anà armar**, eyl e son cavayl.

(Desc III 150, 22) E aytantost l'almugàver trasc son couteyl, e correc sobre·l cavaler qui fo caüt ab lo cavayl, e **anà-li deslassar** son elme e volc-lo degolar; mas lo príncep correc là, e vedà-li-o, e dix-li que son plet avia gasayat e que·l lexàs estar.

(Desc IV 20, 10) E·ls catalans, qui les viuren venir, qui no·s cuydaven que venguessen tro al jorn, bateren de rems e **anaren-se mesclar** ab eles.

(Desc IV 53, 7) E la galea en què·l príncep era, **anà envestir** la galea del almirayl del rey d'Aragó de lonc en lonc, e combateren-se sí fortment, que

(Desc IV 67, 16) E quant lo rey hach açò dit, levà's d'aquí e **anà menjar**

(Desc IV 107, 1) **anaren-se dinar** e

(Desc IV 155, 29; 156, 1) Ab tant, quant de les atzembles dels franssesos hach passades tro a ·C· e dels servents bé ·M·, lo compta **anà desrengar** ab sos ·VII· cavallers; e isqueren del aguayt cridant tuyt: - Aragó! Aragó! -, e **anaren ferir** ardidament sobra·ls fransesos.

(Desc IV 159, 3) E puys **anaren pendra** les atzemblas qui éran romases dels franssesos, e trobàran-hi, entra unas e altres, ·DC·XXII· bèstias;

(Desc V 22, 2–3) E axí com lo rey e los altres hagueren açò hoÿt, lexaren star lo concell, e **anaren-se guarnir** a cavall e a peu, e isquéran a una companya de franssesos qui éran riba d'un stany, prop la vila de Castalló

(Desc V 84, 27–28) que axí com vengueren a la primera juncta, lo rey **anà ferir** ·I· senya-ler franssès

(Desc V 99, 24) e donà tal colp al scuder que tenia lo vantall per la mà, que ultra passà la treta e puys **anà ferir** lo compta qui bevia lo brou e donà-li tal colp per los pits que ultra passaren les enpenes de l'altra part, sí que sempra cahech en sobines mort fret.

(Desc V 109, 1) e ab tant **anà's dinar**

(Desc V 117, 8) Ab tant **anà** ell **ferir** tot primer ab la sua galera, e ferí de tal virtut ab la proha en lo lats d'una galera dels prohansals, que...

(Desc V 145, 10) E los misatgers del rey d'Aragó **anaren dir** la misatgeria al senescal de Tolosa

(Desc II 13, 22) E **van ferir** en la host;

(Desc II 43, 6; 6–7) E el rey, qui assò hac entès, **va pendre** ses armes, e muntà a caval e comensà a córrer aprés d'éls;

(Desc II 113, 10) e els serraÿns qui·ls viren venir aparalaren-se e **van-se mesclar** ab éls.

(Desc II 144, 8) En Bernat Guilem d'Entença, ab los ·L· cavalers e ab los ·M· sirvens, **va ferir** entre·ls serraÿns molt ardidament

(Desc II 144, 21) E **van ferir** en la host dels sarraÿns de la banda de tresmuntana, sí que·ls sarraÿns se desbarataren e comensaren a fugir, en tal guysa que los uns quaïyen morts sobre·ls altres.

(Desc II 178, 30) e, quax qui **va fer** leya ho **sercar** alre, partí's de la barcha

(Desc III 72, 15–16) e aquests malvats ribauts **van-se acostar** a les dones e metien lurs mans a les mameles de les dones.

(Desc III 94, 2–3) E quant fo defora, puyí lo cavayl dels esperons e anà ferir entre·ls sarraÿns, sí que·l primer colp n'abaté, ab lo pits del caval seu, ·III· a terra; e anà ferir ·I· sarraý de la lança, que la dàraqua li passà e tot lo cors, sí que l'abaté mort a terra. E puys arancà-li la lança del cors e, de manén, **va'n donar** tal colp per mig lo pits a aquel qui portava la seyera, que mantinent l'abaté mort a terra.

(Desc III 94, 12) e puyen lurs cavals dels esperons, e **van ferir** en la pressa dels sarraÿns per ajuda al comte de Palars.

(Desc III 120, 7) E aqueles de Marçela, qui estaven de la banda de migjorn, **van dressar** ab gran gatzara ·I·ª seyera molt gran que apelen l'estandart de sen Victor,

(Desc III 121, 15) Ab tant les galeres dels catalans foren aparelades e comensaren a ferir
e·l mig loc de les galeres, là hon aqueles de Piza eren; e quant los marcelesos les viuren
ferir, **van metre** l'estandart de sen Víctor a bayx molt vilment, e preseren la volta de
migjorn, e comensaren de fugir a rems e a vela,

(Desc III 123, 24) Quant les gens de Messina saberen aquela nuyt la novela, levaren-se
tuyt de lurs lits, grans e pocs, e **van encendre** tortes, e brandons, e cera, e candeles e
fayes

(Desc III 139, 3) – A armes! a armes, cavalers francès! ¡Que les gens del rey d'Aragó són
ab nós! – E cascú **se va armar** axí com poc, e d'altres qui fugien sens armes;

(Desc III 141, 2) Puis pres ·I· salt a través, e encontrà ·I· cavaler d'aqueyls ·V·, e a través
per mig los flancs **va ferir** de la lança lo cavayl; sí que de l'altra part li'n passà tot lo
ferre, e·l cavayl caec mort a terra e el cavaler caec sots lo cavayl, sí que no·s pudia levar.

(Desc III 160, 9) Ab tant lo senescalc **se va** a eyl **acostar**, e saludà-lo e dix-li que ben fos
vengut;

(Desc IV 87, 20–21) Quant açò hach dit lo rey, vench lo mestre ab son pich e ab son
capmartell, e **va obrir e trencar** loses de la cambra qui era losade,

(Desc IV 88, 5) E puys lo mestra **va entrar** dins ab lums de candeles

(Desc IV 123, 23) e quant foren ben prop, aquells laïns **van desserrar** les ballestes, peyras
e cantals a gitar per los murs e matéran grans crits e grans veus, que paria que n'i
hagués ·X· tants més que no·y havia

(Desc IV 124, 3) e **van trencar** les portes del monestir e barrajaren e robaren la sglésia e
totes quantes coses hi hach

(Desc IV 152, 26–27) ·I· d'aquells qui staven al mur del castel dix a ·I· ballaster que li
stava de prop, que li tiràs e que no·l escoltàs pus; e lo ballaster **va desparar** la ballesta. E lo rey, que·u viu, punyí son cavall dels sperons, e tornà-sse'n a sa companyia e
tench-sa molt per scarnit d'aquest fet.

(Desc V 16, 23–4) E quant lo compte de Pallàs e ceis qui éran dins la vila de Peralada
viren aquella companya que axí se era acostada al mur, guarniren-sa tuyt, cavallers e
servents, e éren tro a ·CL· cavallers armats e ben ·III·ᵐ servents, e **van obrir** les portes
de la vila, e donaren tuyt salt defora ardidament, e feriren en aquella companya de
franssesos.

(Desc V 99, 20) E quant lo serraý l'hach bé aesmat, **va desparar** sa bona ballesta de ·II·
peus que tenia

(Desc V 101, 4–5; 8) cuydaren-sa que aquells de laïns se'n fossen fuyts amagadament la
nit traspassada e **van pujar** ab gran goyg per les scales amunt. E quant n'ach pujats en
les scales de ·CCC·L· en ·CCCC· e foren ja bé al mig loch, En Ramon Folch féu tocar
l'anafil, e los de llaïns **van gitar** les labreres demunt dites per les escales avall e levaren-
les axí e buydaren-les, que tots quants hi éran pujats, tots anaren a terra

(Desc V 116, 17–18) Ab tant les galeres del rey d'Aragó **van tocar** les trompes e los tabals
e cridaren a grans crits:

Anar a + Inf

(Desc II 59, 18) Com la emperadriu de Alamanya **anà a sercar** lo bon comte de Barcelona

(Desc III 10, 4) Com lo rey de Castella **anà a parlar** ab lo apostoli

(Munt)

Anar + Inf

(Munt I 33) E així mateix **anà visitar** Montpestller, de què ell havia gran desig a visitar.

(Munt I 35) Sí que hac sos missatges bons e honrats qui **anaren fermar** lo fet ab los missatges del rei Manfré que per aquella raó eren venguts;

(Munt I 37) E tantost lo dit senyor rei de Castella llevà's e **anà besar** en la boca lo dit senyor rei En Jacme e dix-li:

(Munt I 61) E con les dites galees foren armades e el dit En Corral se volc recollir, **anà pendre** comiat del senyor rei, qui era en la ciutat de Lleida.

(Munt I 62) E així, lo dit Corral ab les cartes e capítols partí's del senyor e **anà's recollir** en València en les cinc galees. E puis anà a Barcelona,

(Munt I 65) E lo papa llavors **anà'l besar** en la boca, e dix-li:

(Munt I 84) E llevà's e **anà pendre** comiat de madona la reina a dels infants

(Munt I 102) E no bastà açò, que enans **anaren combatre** Nicòtena, e preseren-la;

(Munt I 124) E sobre açò En Guillem Cornut llevà's, e **anà besar** lo peu al rei Carles, e dix:

(Munt I 137) E En Domingo, qui oí açò, llevà's, e **anà's ajonollar** al senyor rei, e besà-li lo peu, e dix:

(Munt I 141) e con hagren sopat, **anaren-se gitar**.

(Munt I 143) e **anà aorar** a la capella.

(Munt I 145) sí que la major valor e cortesia féu que jo anc veés fer a senyor: que comiat **anà pendre** de la dona muller sua, e hi avallà, e hi muntà en les cambres, així con si fos en lo pus segur lloc del món.

(Munt I 152) E les trompes tocaren; e **anaren menjar**.

(Munt I 153) e així són tres persones que podets comparar als tres Reis qui **anaren aorar**, los tres Reis, Jesucrist

(Munt I 154) E les trompes e les nàcares tocaren, e **anaren menjar**; e tothom qui menjar volgués, poc menjar.

(Munt I 157) E tantost con hac jurat, llevà's e **anà besar** la mà a madona la reina e als infants;

(Munt I 158) e **anà's veure** ab son nebot lo rei de Castella

(Munt I 166) E així, con tot açò fo fet, l'almirall avallà a la Duquena, e entrà-se'n al palau e **anà fer** reverència a madona la reina;

(Munt I 166) E així mateix, con hac feta reverència a madona la reina, **anà besar** la mà a dona Bella, sa mare;

(Munt I 167) E con açò fo otorgat, lo senyor infant e son consell **anaren tenir** llur consell lla on havien acostumat, de tots fets;

(Munt I 170) e el senyor infant dix-li ço que havia pensat, e que pensàs de muntar en les galees, e que pensàs de ben a fer. E En Berenguer de Vilaragut **anà-li besar** la man e li féu moltes gràcies;

(Munt I 171) E con açò fo fet, **anà visitar** Saragossa, e Not e tota la vall de Not;

(Munt I 171) E con ell veé que negun no li volia eixir, una nit partís de Brandis e **anà barrejar** Vilanova, e puis Pollina, e puis tot lo burg de Manòpol. E con tot açò hac barrejat e en cascun lloc hac pres molt navili, lo qual ne tramès a Messina, **anà córrer** la illa de Curfó;

(Munt I 175) E ab aitant besaren-li la mà e **anaren fer** ço que el senyor rei los hac manat.

(Munt I 177) E con foren recollits, l'almirall **anà pendre** comiat de madona la reina e dels infants.

(Munt I 181) E con foren al peu de l'escala, madona la reina e sa germana **anaren-se abraçar**; e estegren així abraçades, e besant e plorant,

(Munt I 183) e us ha aportat en poder la pus cara cosa que Carles havia e'l món, pensats-lo de jutjar, e dats-li aquella sentència que us parrà que sia vista. – E sobre açò **anà seure**.

(Munt I 184) E sobre açò llevaren-se e **anaren menjar**

(Munt I 198) Què us diré? Ella va metre la man a l'espaa, e venc a un portell altre, e **anà ferir** lo cavall per la testera, e el cavall estec estabornit.

(Munt I 202) E con açò fo fet, lo rei de França **anà's atendar** a Gerona, e posà son setge.

(Munt I 213) e lo senyor rei baixà's, e **anà'ls abraçar**, e reebé'ls ab bella cara e ab bell semblant.

(Munt I 216) per què ells, ab llicència del rei de França, **anaren-lo çercar**.

(Munt I 227) E tantost l'almirall pres comiat del senyor rei, e **anà's recollir**, e féu la via de Salou ab totes les galees;

(Munt I 229) E lo senyor infant ab tota la cavalleria e l'almogaveria **anaren-se'n atendar** a les Torres Llavaneres;

(Munt II 14) E sobre açò lo senyor rei N'Anfòs pres per la man lo senyor infant En Pere, qui li seïa de prop, un poc pus baix, e **anà'l besar**, e dix-li: - Infant, aital resposta esperàvem de vós, e aital fe hi havíem.

(Munt II 18) E puis partí d'aquí, e **anà barrejar** la ciutat de Patraix.

(Munt II 27) e així los almogàvers aquells **anaren percaçar** cols, peix e fruita que menjassen.

(Munt II 29) E lo senyor rei féu llevar lo camp a les sues gents, e puis **anà assetjar** los castell on lo moixerif s'era més.

(Munt II 44) E el solaç e el deport se moc entre ells, que el senyor rei d'Aragon **anà veure** lo rei Carles e la infanta muller sua;

(Munt II 48) E con ja us he dit, lo senyor rei d'Aragon **anà** una vegada **veure** lo papa en Roma,

(Munt II 66) E el duc **anà desemparar** Catània e tots los altres llocs que tenien en Sicília;

(Munt II 64) E de Térmens partiren, e **anaren assetjar** la vila de Xaca,

(Munt II 75) e tothom llevà's e **anaren-li besar** la mà, e li faeren moltes gràcies.

(Munt II 92) e aquell, qui es veé afollat, qui era macip ben temprat, **anà'l abraçar**, e ab una brotxa donà-li ben tretze coutellades,

(Munt II 92) així que **anà ferir**, en una entrada que féu, un mariner per nom Bernat Ferrer, qui era sobre un bon cavall que havia guanyat en la batalla primera...

(Munt II 93) E con açò hagueren fet, **anaren pendre** altra ciutat,

(Munt II 94) e **aná córrer** estrò prés de la ciutat de Contastinoble.

(Munt II 94) E així guanyà tant, ell e sa companya, en aquella cavalcada, que ab aquell guany **anà assetjar** un castell qui és a l'entrant de Boca d'Aver,

(Munt II 94) e **anà's acostar** al mur, e féu semblant que acostàs escales, e null hom no hi parec.

(Munt II 98) E el cavaller, qui veé que l'aconseguien e que la dona s'havia a perdre, brocà un poc avant, e la dona gità-li un gran crit, e ell tornà envers ella e **anà-la abraçar e besar**; e con ho hac fet, donà-li tal de l'espaa per lo coll, que el cap ne llevà en un colp.

(Munt II 101) e a la fin deu galees **anaren pendre** terra lluny. E sobre aquell pendre de terra, a mi caec lo cavall, e finalment un meu escuder avallà e donà-m lo seu;

(Munt II 105) Què us diré? Que En Rocafort, ab los turcs e ab l'almogaveria la major part, **anà tenir** setge a la ciutat de Nova, qui era a seixanta milles lluny de Gal·lípol. E En Berenguer d'Entença **anà tenir** setge a un castell qui ha nom lo Megareig,

(Munt II 119) E així partí'm d'ell, e haguí pres comiat del senyor infant e de sa companya, e **anè pendre** comiat del duc; e ell, la sua mercè, donà'm ses joies riques e bones, e partí'm pagat d'ell.

(Munt II 123) E el comte dix que anassen a la mala ventura, que bo era que morissen ab los altres. E així tots plegats **anaren-se mesclar** ab la companya. E pensaren tuit d'arrengar batalla.

(Munt II 123) E ell mès-se en la davantera ab la sua senyera, e pensà de brocar, e va ferir en la companya; e aquells de la companya **anaren ferir** en ell. Què us diré?

(Munt II 130) e tractà ab lo rei de Castella que de tot anassen sobre el rei de Granada en aquesta manera: que el rei de Castella ab son poder anàs assetjar Altzehira d'Alhadre, e lo senyor rei En Jacme d'Aragon anàs assetjar la ciutat d'Almeria. (…) E així es complí, que el rei de Castella **anà assetjar** Altzehira d'Alhadre e el senyor rei d'Aragon la ciutat d'Almeria,

(Munt II 132) en aquella conquesta, qui sua no era, **anà tenir** setge;

(Munt II 136) E així, com los sarraïns veeren envers ells venir los crestians sens neguna ordinació, **anaren ferir** en los primers.

(Munt II 138) E **ané-li besar** la mà, e la li'n besaren molts rics-hòmens e cavallers per mi.

(Munt II 137) E jo, qui oí que el senyor rei havia tan gran fe en mi en aquests afers, vaig-me senyar, e **ané'm agenollar** davant ell

(Munt II 147) e puis per mar e per terra **anà assetjar** Tràpena.

(Munt II 155) E con fo a Messina, lo senyor rei e el senyor infant **anaren-les veure**;

(Munt II 158) E jo llevé'm, e **ané-li besar** la mà

(Munt II 159) E tantost jo me llevé, e **ané-li besar** la mà

(Munt II 160) E tantost con aquells de la ciutat de Clarença l'hagueren jurat, ell **anà assetjar** los castell de Belveer, qui és un dels bells castells del món, e és prop a Clarença; e combaté-lo fortment,

(Munt II 185) E en cascuns dels llocs pres gran damnatge, sí que de Catània s'hac a llevar, e **anà's recollir** a l'estol seu a la terra de Màscara; e d'aquí partent, pres terra a Cobogròs, qui és prop de Messina quinze milles;

(Munt II 211) **anà siure** davant l'altar de sent Salvador, al siti reial

(Munt II 89) E jo tantost e tots los altres **anam fer esfondrar** tots los vaixells.

(Munt II 90) e així prenguem vigoria, e **anam** tuit **ferir** fermament en ells, e així mateix venceren-se;

(Munt II 90–91) E **anam llevar** lo camp, e segurament que atrobam que tota hora haguem morts ben sis mília hòmens a cavall e més de vint mília de peu.

(Munt II 102) jo em comané a Déu e a madona Santa Maria, e vaig fer obrir la porta. E ab los sis cavalls armats e los hòmens de peu que eixiren així lleugers, **anam ferir** en les senyeres, sí que al primer colp n'abatem tres.

(Munt II 119) E pensam de partir de Negrepont, e **anam refrescar** a la illa de Setepose, e puis a la Sidra, e puis a…

(Munt II 143) e aquell jorn **anam-nos atendar** prop d'ells a mija llegua.

(Munt I 47) E ab aitant féu tocar les trompes e les nàcares, e ab gran brogit e ab grans crits començaren a fer un baixa mà molt vigorós. E les quatre galees tot bellament e menys de crits e de paraules e de tabustol negun, **van ferir** enmig de les deu galees.

(Munt I 91–92) Que con lo rei veé que ells rebujaven e s'anaven aturant, cuità's e manà la davantera que ferís, sí que féu desplegar l'estendard, e les trompes e les nàcares tocaren, e la davantera **va ferir**. Sí que els sarraïns tengren molt fort, així que els crestians no els podien esvair; tanta era la gent. E lo senyor rei brocà ab la senyera e **va ferir** entre els moros, e los moros se desbarataren en tal manera que no en romàs negun qui en la davantera dels sarraïns fos qui n'escapàs,

(Munt I 97) que així són acostumats e nodrits: que con **van fer cavalcada**, cascun porta un pa per cascun dia, e no pus; e puis del pa e de l'aiga e de les herbes, passen llur temps aitant con ops llurs és.

(Munt I 101–102) E tantost con lo senyor rei fo en terra, les galees **van batre** de rems;

(Munt I 102) així que, totes vint-e-dues enfrenellades, e embarbotades, **van rogar**, aparellats de la batalla, contra l'estol del rei Carles. E aquells de l'estol del rei Carles no es podien pensar, per res, que haguessen cor que es combatessen, mas que se'n faessen afaenats. Mas, con veeren que a de veres se faïa, les deu galees de pisans **van eixir** de l'esquerra e **van arborar**, e en redon, ab lo vent que era fresc, **van-se metre** en mar e pensaren de fugir. Con los pisans hagueren açò fet, semblantment faeren los genoveses, e així mateix los proençals,

(Munt I 102) E con les quaranta-e-cinc galees, e los llenys armats, e les barques de Principat veeren açò, tengren-se per morts, e **van ferir** en la plaja de Nicòtena. E les vint-e-dues galees **van ferir** entre ells.

(Munt I 103) e tantost **va pendre** terra

(Munt I 111) – Doncs, rei, pus així ho deïts, dats-nos vostre gatge en presència de tuit. E llavors lo rei pres un parell de guants que tenia un cavaller, e **va-los gitar** en presència de tuit. E los missatges del rei Carles preseren los gatges,

(Munt I 126) e tantost, ell en persona, ab tres galees, acostà's a ells. E ells, qui el sentiren, **van-se pendre** a rems, que més se fiaven en los rems que en Déus ne en les armes. E l'almirall **va-les ferir**.

(Munt I 128) E enmig del port **van-se ferir** tan vigorosament, que totes les proes de cascuns se romperen;

(Munt I 128) e tuit començaren a pendre vigoria e **van muntar** en les galees dels proençals, e tots quants ne trobaren sobre coberta mataren.

(Munt I 129) E així **van pendre** totes les vint-e-dues galees

(Munt I 129) E con l'almirall En Roger de Lória hac preses les galees e el lleny, anà-se'n a la punta del port, de ponent, e **va metre** la sua gent en terra;

(Munt I 143) E llavors ell lo guardà, e conec que era el rei d'Aragon. E **va donar** del peu en terra, e volc-li besar lo peu;

(Munt I 143) E llavors ell **va brocar** lo cavall, e environà tot lo camp

(Munt I 151) E con açò fo fet, lo senyor rei de Mallorca, e comtes, barons, e prelats, cavallers e ciutadans, tots **van eixir** del palau.

(Munt I 164) E així, aparellats de la batalla, **van-se venir**.

(Munt I 164) E l'almirall e les sues galees pensaren de pendre vigoria, e cridaren: – Aragó! Aragó! Via sus! Via sus! –. E **van pujar** en les galees tothom;

(Munt I 179) E con açò hagueren cascuns fet, **van-se ferir** les unes galees ab les altres;

(Munt I 179) E sobre açò, l'almirall mès força contra força, e poder contra poder, e cridà: – Vergonya! Vergonya! –. Sí que tothom pensà de saltar en la galea del príncep, e **van escombrar** tota la proa;

(Munt I 192) E així con ells se degren humiliar a les creus, ells **van ferir** en ells e tallaren-los tots, e clergues, e llecs, e hòmens, e fembres e infants, per la fellonia d'açò que els era esdevengut.

(Munt I 197) E el senyor infant, escalfat de bona cavalleria, volc fer brocar la senyera per anar ferir en ells, mas lo comte de Pallars ne els altres no li ho consentirien. Què us diré? Que envides lo podien cabdellar que no ferís, entrò que el comte de Pallars lo **va pendre** per lo fre e dix:

(Munt I 198) Què us diré? Ella **va metre** la man a l'espaa, e venc a un portell altre, e anà ferir lo cavall per la testera, e el cavall estec estabornit. Què us diré? Lo cavall **va pendre** per la regna, e cridà: – Cavaller: mort sóts, si no us retrets. – E el cavaller tenc-se per mort, e **va tenir** la bordó e **va'l gitar** en terra, e retés a ella;

(Munt I 199) ells **van metre** bé en cent parts de la vila de Peralada foc, e cridaren: – Eixir! eixir!

(Munt I 199–200) E així vengren a l'esgleia e atrobaren aquella bona fembra qui tenia abraçada la image de madona santa Maria; e **van venir** los malvats de picards, qui era la pijor gent de la host, e especejaren tota la bona fembra sobre l'altar;

(Munt I 210) E con açò hagren vist, **van fer tocar** la trompeta de la llur galea, qui era senyal emprès que

(Munt I 211) e **van-los dir** aquelles novelles

(Munt I 214) e los francesos qui estaven emboscats, pensaren-se que fossen descoberts, e tantost **van eixir** de la celada.

(Munt I 214) e **van migpartir** les llances. Lo senyor rei **va brocar** primer, e **va ferir** lo primer que trobà ab la llança tal per mig de l'escut, que no li calc cercar metge; e puis **va metre** mà a l'espaa, e donà deçà e dellà, que tal lloc se faïa fer, que negun no el gosà esperar a dret colp, con l'hagueren regonegut del seu ferir.

(Munt I 215) E estant en aquella pressa, un cavaller francès veé que el rei los faïa tant de damnatge, ab l'espaa en la mà ell **va venir** e tallà-li les regnes; sí que per açò lo rei se cuidà perdre.

(Munt I 216) Que els francesos se volgren replegar en un cabeçó, mas lo senyor rei **va brocar** envers aquell qui portava la senyera del comte, e **va**-li tal **donar** de la maça de sus en l'elm, que mort fred lo mès en terra; e los almogàvers **van** tantost **trer** la senyera de l'asta, a trosses. E los francesos, qui veeren la senyera de llur senyor en terra, **van fer** tots una mota, e lo senyor rei **va ferir** entre ells ab tots los seus ensems.

(Munt I 218) e **va fer recúller** la gent, e pres comiat de madona la reina e dels infants.

(Munt I 219) e con fo ab ells, a colp los fanars foren encesos, e **van ferir** en ells. E aquí veérets llances e dards volar e ballesters en taula jugar.

(Munt I 219) e con se foren acostats, l'almirall met rems, e **va fer abatre** aquelles senyeres, e mès aquelles del senyor rei d'Aragon

(Munt I 220) E aquelles, qui veeren açò, volgren girar, mas l'almirall los **va ferir**. Què us diré? Que totes las hac, ab tota la gent.

(Munt I 220) e **van ferir** en ells, e desconfí-los, sí que hi mataren més de dos-cents cava- llers franceses e moltes asembles qui hi eren.

(Munt I 224) Lo rei de França s'acostà al rei de Mallorca, e el rei de Mallorca a ell, e **van-se besar e abraçar**;

(Munt II 9) e con foren prop, que els dards pogren jugar e les ballestes a colp trer a triquet, les trompes e les nàcares **van sonar**, e l'almirall, ab los hòmens de cavall, **va brocar** als hòmens de cavall, qui eren ben tres-cents d'altra part, qui franceses, qui del país. E los almogàvers, qui hi havia ben dos mília, **van tremetre** los dards que anc un no en pecà que no metés hom mort, o nafrat per mort; e los ballesters a colp **van desparar**. Així que tant fo l'esforç al començament que l'almirall e sa companya faeren, que cridaren: - Aragó, Aragó! -, que a colp aquells **van girar**,

(Munt II 10) Mas con foren prop d'Agda, noves los **van venir** con havia pres, lo dia passat, a aquells de Besers; e així con ho oïren, pensaren de girar;

(Munt II 18) Emperò comanà's a Déu, e féu replegar sa gent, e **va ferir** en ells tan esforçadament, que per cert que els féu tornar a enrera, envers la ciutat (…) E los almogàvers, qui veeren aquesta pressa e que els francesos se tenien tan fort, **van trosse- jar** llances, e puis **van-se metre** entre ells, e pensaren d'esbutllar cavalls e de matar cavallers.

(Munt II 27) e mès sa tovalla, e tuit començaren a riure e aixufar, que cuidaven-se que ho fes per escarn. E con ells foren asseguts, començaren a menjar, e aquell pres son quarter de moltó, e **va'l-se posar** davant, e tallà de la carn, e dix:

(Munt II 29) e con les hosts foren la una prop de l'altra, lo senyor rei **va ferir** ordonadament ab tota la sua gent, e el moixerif així mateix contra lo senyor rei. E la batalla fo molt cruel,

(Munt II 49) e lo dit En Berenguer de Puigmoltó saltà avant, e un altre cavaller. E aquells de dins **van-se** aquí **defendre**, així que per cert aquells dos cavallers foren morts si no fos lo senyor rei,

(Munt II 50) E lo senyor rei saltà dins, avant, qui era jove e trempat, e **va**-li tal **donar** per mig del cap, de l'espaa, que el capmall que portava no li valc un diner, que entrò en les dents lo fenè; e puis **va-li trer** l'espaa del cap, e **va'n ferir** altre, que el braç ab tot lo muscle n'avallà en terra.

(Munt II 53) E con cascuna de les hosts se veeren, los almogàvers del comte Galceran e de don Blasco cridaren: -Desperta, ferres! Desperta!-; e tots a colp **van ferir** dels ferres de les llances en les pedres, sí que el foc ne faïa cascun eixir, així que paria que tot lo món fos llumenària, e majorment con era alba.

(Munt II 53) e cridaren: -Aragó! Aragó! -; e llavors aquell nom escalfà'ls tots, e **van** tan règeu **ferir**, que açò fo la major meravella del món (…) e llavors lo comte Galceran e don Blasco, rompent, **van ferir** en ells.

(Munt II 53) E llavors senyà's e comanà's a Déu, e, batalla arrengada, **van-se ferir** los uns e'ls altres.

(Munt II 53) E lo comte Galceran e don Blasco no volgren fer davantera ni saguera, ans tots plegats, la cavalleria de la banda sinestra e los almogàvers de la dreta, **van ferir** la davantera d'aquells, en tal manera que parec que tot lo món ne vingués. E la batalla fo molt cruel, e los almogàvers **van trametre** los dards que endiablia fo ço que ells ne faeren; (…) E puis **van trossejar** les llances e **esbutllar** cavalls, que així anaven entre ells con si anassen per un bell jardí. E lo comte Galceran e don Blasco **van-se faixar** ab les senyeres dels franceses, en tal manera que a terra les gitaren totes.

(Munt II 55) E així, les senyeres de les davanteres de cascunes parts s'acostaren, e **van-se ferir** tan fortment, que açò fou una gran meravella.

(Munt II 55–56) no volc pus esperar, ans se n'anà tot dret lla on viu la senyera del príncep, e **va ferir** tan vigorosament, que ell, son cos, donà tal de la llança al banderer del príncep, que en terra mès ell e la bandera.

(Munt II 56) E a la fin lo senyor rei donà tal de la maça al cap del cavall del príncep, que el cavall fo fora de tot son seny e **va caure** en terra;

(Munt II 65) e el jorn que la vista fo ordonada, foren cascuns d'ells a la vista, e **van-se besar e abraçar.**

(Munt II 67) E sobre açò frare Roger **va besar** la mà al senyor rei, e partí's d'ell;

(Munt II 70) E con foren prop los uns dels altres, los trenta escuders **van brocar** e **van** tal **ferir** lla on era la senyera dels genoveses, que abateren a terra aquell Rosso de Finar, e los almogàvers **van ferir** en ells.

(Munt II 76) E la companya hac-ne gran plaer, sí que aitantost, abans que soferissen les fletxes dels arcs dels turcs, **van brocar** en ells, los de cavall a aquells de cavall, e los almogàvers aquells de peu.

(Munt II 76) e la companya pres-se a les armes e pensà's de cuitar. E En Corberan d'Alet, entrò ab dos-cents hòmens a cavall e mil de peu, **va** entre ells **ferir**, sí que tantost los mès en veençó.

(Munt II 77) e los turcs, qui veeren que els muntaven darrera, pensaren ab les sagetes de trer, e per desastre una sageta **va ferir** En Corberan, qui s'hac desarmat lo cap per calor, per lo pols. E aquí ell morí, de què fo gran tala,

(Munt II 78) E la host eixí en tal manera que aconseguiren los turcs e **van ferir** en ells, sí que en aquell jorn mataren ben mil hòmens de cavall, de turcs, e ben dos mília hòmens de peu.

(Munt II 79) e tantost lo megaduc ab la cavalleria **va ferir** als hòmens a cavall, e En Roca-fort ab l'almogaveria als hòmens de peu, e aquí veérets fets d'armes, que jamés tal cosa no veé null hom.

(Munt II 85) E aquell dia ell convidà lo cèsar e con hagren menjat, aquell Girgon, cap dels alans, entrà en lo palau on estava xor Miqueli ab sa muller e el cèsar, e **van trer** les espaes e **van** tot **especejar** lo cèsar e tots aquells qui ab ell eren;

(Munt II 85) E tantost **van-nos posar** setge davant, e venc tanta de gent sobre nós, que foren més de deu mília hòmens a cavall

(Munt II 89) e així con la senyera se llevà, un núul se mès sobre nós, e **va'ns cobrir** tots d'aigua així con estàvem agenollats, e durà aitant com la "*Salve Regina*" durà a cantar.

(Munt II 90) per pecat llur e per lo bon dret que nós havíem, **van-se vençre**;

(Munt II 94) E En Ferran Xemenis, qui els veé, preïcà la sua gent e els amonestà de bé a fer, e tuit ensems **van ferir**. Què us diré?

(Munt II 95) E **van arborar** les escales, sí que quatre escales ab los rampagolls meteren e'l mur; e puis **van muntar** en cascuna escala cinc hòmens, un aprés altre, e tot suau muntaren-se'n al mur, que anc no foren sentits; e pus muntaren-hi altres vint, e així foren quaranta, e **van-se emparar** de dues torres.

(Munt II 98) E con açò hac fet, tornà sobre les nostres tres hòmens a cavall, qui ja preien lo cavall de la dona, e ab l'espaa **va** tal colp **donar** a un d'aquells, qui havia nom Guillem de Bellveer, que el braç sinestre li n'avallà en un colp, e caec en terra mort.

(Munt II 98) e puis llevaren-se matí, e a alba foren ab ells e **van ferir** per les tendes.

(Munt II 102) E jo llavors mané a un meu escuder, qui era a cavall armat, que brocàs sobre ell; e aquell féu-ho volenter, e **va-li** tal colp **donar** del pits del cavall, que en terra lo mès, e llavors foren-ne fetes cent peces.

(Munt II 102) jo em comané a Déu e a madona Santa Maria, e **vaig fer obrir** la porta. E ab los sis cavalls armats e los hòmens de peu que eixiren així lleugers, anam ferir en les senyeres, sí que al primer colp n'abatem tres. E ells, qui hagren vist que nós així ferim vigorosament, de cavall e de peu, **van-se vençre**, així que tantost n'haguem les espatlles.

(Munt II 110) A con l'hagren mort, **van cercar** los altres, e assenyaladament En Ferran Xemenis. E En Ferran Xemenis, així con a bon cavaller e savi, així mateix al brogit, muntà en son cavall, tot desguarnit, e anava capdellant.

(Munt II 110) e a junta **van venir** a En Berenguer d'Entença, qui cabdellava, e cui-daren-se que ell enegàs la companya. E abdosos a junta **van-li venir**, e En Berenguer d'Entença cridà, e dix què s'era; e ells no guardaren àls, mas abdosos lo **van ferir**. E trobaren-lo desarmat, e passaren-li tres llances de l'altra part, així que abdosos aquí mateix lo mataren;

(Munt II 115–116) E con fom en terra, les galees dels venecians **van córrer** sobre les nostres, e senyaladament sobre la mia, cor era veu que jo traïa de Romania tot lo tresor del món. E al muntar que hi faeren, mataren-me ben quaranta hòmens, e així mateix hagren mort mi si hi fos, mas jo no em partia un pas del senyor infant.

(Munt II 116) E com açò fo fet, tragueren mi en terra, e con aquells de la companya me veeren, e En Rocafort e els altres **van-me** tots **besar e abraçar**, e començaren tuit a plorar d'açò que havia perdut.

(Munt II 121–122) E moc-se entre ells una gran rumor, que **es van pendre** a les armes, e **van allancejar** catorze caps de companya qui en aquell fet havien consentit. E puis **van eléger** dos de cavall, e un adalill e un mogatèn, per qui es regissen estrò haguessen cap.

(Munt II 123) E ell mès-se en la davantera ab la sua senyera, e pensà de brocar, e **va ferir** en la companya; e aquells de la companya anaren ferir en ell. Què us diré?

(Munt II 128) E misser Bonifaci, qui veé que ab cor d'atendre ho deïa, acostà's a l'altar, a l'arquebisbe, e aquí ell **va fer** lo duc cavaller.

(Munt II 131) e **va brocar** sobre ell; e abans que a ell se pogués acostar, hac morts ab la llança més de sis cavallers, e hac rota la llança. E puis **va metre** la mà a l'espaa, e ab l'espaa en la mà ell se féu fer lloc tant estrò que venc a aquell qui cridava que era fill del rei; e aquell, qui el veé venir, sabé que era l'infant, venc envers ell, e **va-li**, de l'espaa, **donar** tal colp, que el primer quarter de l'escut la gità en terra; e fo molt meravellós colp. E cridà: - Ani ben e soltan!- E lo senyor infant **va**-li tal **donar** de l'espaa per lo cap, que entrò a les dents lo fené; e caec mort en terra.

(Munt II 137) E jo, qui oí que el senyor rei havia tan gran fe en mi en aquests afers, **vaig-me senyar**, e ané'm agenollar davant ell

(Munt II 141) E així con vengren e fore prop de mi, jo **vaig ferir** enmig d'ells en tal manera, que set de les barques mis a fons, e **vaig-me'n calar** sobre elles; e pensam de ferir sobre ells, deçà e dellà, ab los altres llenys e barques, qui tantost feriren.

(Munt II 143–144) e així cuitaren-se, e **van-se llevar** ensems, e **van ferir** en la nostra peonada, en tal manera que els metien endarrera.

(Munt II 144) e aprés, misser Corral e tots los altres **van ferir**, que anc lo terç senyal no pògrem fer.

(Munt II 148) E tantost **van-me fer** sagrament tots, e em besaren en la boca

(Munt II 159) Què us diré? Que los almogàvers qui eren ab ell, **van eixir** en terra, e els ballesters, e **van ferir** en aquella gent, sí que los faeren llunyar e fer plaça;

(Munt II 160, 160, 160) llavors féu desplegar la senyera, que pus no volc esperar l'altra cavalleria sua, ans féu sa brocada ab aquells cinquanta hòmens a cavall e ab los almogàvers; e **va ferir** als enemics, e **va'ls desbaratar**, així que **van fugir** envers la ciutat, e el senyor infant e sa companya ab ells firent e donant. Què us diré?

(Munt II 175–176) Con les hosts foren apropriades, cascuns **van brocar** molt vigorosament, sí que jamés no pògrets veer batalla pus cruel

(Munt II 176) E així ho faeren, que tots set ensems brocaren, e **van-li matar** lo cavall

(Munt II 176) e **va trer** lo bordó que portava cint; e veé la sua senyera en terra, e ab lo bordó en la mà, **va llevar** la sua senyera, a la dreçà e la tenc abraçada. E sobre açò, un cavaller seu, per nom En Bernat de Boixadors, **va avallar** del cavall e **va pendre** la senyera, e lliurà lo cavall al senyor infant; e lo senyor infant lo muntà tantost, e féu pendre la senyera a un cavaller. (…) e ab lo pom del bordó al pits, **va brocar** envers ell, e **va**-li tal colp **donar** per mig del pits, que d'altra part lo li passà; e caec en terra mort,

(Munt II 177) Com los almogàvers e sirvents de mainada veeren començar la batalla dels cavallers, estrò a dos-cents **van trencar** per mig les llances, e meteren-se entre los cavalls, a esfondrar; e los altres **van ferir** en la llur peonada, tan fortment que ab los dards ne mès cascun un en terra

(Munt II 177) Que la batalla fo tan fort, que a un colp **se van tirar** tots los tudescs e els pisans qui eren romases, e **van-se emparar** d'un puget;

(Munt II 178) e aparellaren-se de cavall e de peu, e **van eixir**, que aquells del setge no en veeren res.

(Munt II 178) Sí que el senyor infant, qui ho entès, qui dormia en espatlleres, **va's llaçar** lo capell de ferre e **posar** l'escut al coll;

(Munt II 191) e tantost **van vogar**, e abans que no pogueren haver girat aquelles set galees, l'almirall los fo a dors, e pensà'n en tal manera, que més de mil e cent persones hi mataren;

(Munt II 193) que alba de dia tots foren entorn dels murs, e **van-se faixar** als murs tan vigorosament, que no guardaren perill qui esdevenir-los-en pogués;

(Munt II 193) vengren tots d'aquella part, e los hòmens de cavall **van-se acostar** al mur, e **van**-lo així mateix **esvair**.

(Munt II 194) E con açò hagren fet, l'endemà **van-hi tornar** tots, e **van enderrocar** tots los murs e les cases, e ho meteren tot a pla. E los dits nobles ordonaren que la pedra e la fusta pensàs tothom de pendre a llur volentat, e que ho portassen a Bonaire; e tothom ho pensà de carrejar, qui ab barques, qui ab càrrous, e **van-ho tot portar** a Bonaire, e **van-ne bastir e obrar** bons albergs.

Anar a + Inf

(Munt I 198) E així lo vescomte de Rocabertí **anà** per los seus castells **a enfortir**;

(Munt I 135) E estant que els fets s'aparellaven, lo senyor rei **anà** a Saragossa, **a visitar** la ciutat e **veure** madona la reina e els infants.

(Munt II 57) qui així mateix hi **anaren a pregar** lo rei de França que trametés son frare, misser Carles, en Sicília en ajuda del duc;

(Munt II 114) E en aquell castell venguí jo, e aquí trobé lo senyor infant ab les quatre galees, e aquí m'esperà ell con jo **ané** a la companya **a pendre** comiat, e aquí torné al senyor infant.

(Munt II 118) E tantost féu obrir les portes del castell de Sent Omer, on ell estava, e jo **ané a veer** lo senyor infant.

(Per)

Anar + Inf

(Per 45) E los dimonis que aquí m'havien menat se restaren a la riba del fluvi, e quan ells **anaren veure** que io me n'anava així segurmant sobre lo pont, feren molt gran do e molt horrible e tan espaventable,

(Per 34) E io llavores li **vaig dir** que, ab l'ajuda de Déu, io hi intraria per purgar mos pecats;

(Per 35) En recomanant-me a les bones oracions dels bons prohomes, e garní-me de fe e de creença lo millor que jo poguí, e **vaig-me senyar** del de la creu, e me recomané a Déu, e intré dedins lo purgatori, e mon companyò aprés.

(Per 36) e intré dedins e **vaig-me seure**.

(Per 38) E açò ells **me van dir** per me decebre e per menaces e per mentides;

(Per 38) E quan io vi açò, io **vaig recobrar** cor e fui molt pus ardit que davant,

(Per 44) e **vaig-lo reclamar** Jesu Crist e tota sa ajuda per la molt gran angoixa que io hi sentia e per la gran dolor e turment ni eren ni hi són.

(Per 48) E aprés que io haguí oïts los dolces cants e melodies, adoncs los dos arcebisbes que m'havien menat dedins me tiraren a part e **van-me dir**:

(Per 49) E quan ells haguen ab mi parlat així llongament, me menaren en una gran muntanya, e digueren-me que regardàs en aut e en avant, vers lo cel, e io hi **vaig regardar**, e **van-me demanar** de quina color era ni cum me semblava ni ont era io. E io lus **vaig respondre** que a mi me semblava color d'or e d'argent de la fornal ixit. E llavores ells **me van dir**:

(Per 50) Aquí fóra io volenterament aturat si hagués pogut, mes, aprés aquestes coses que a mi foren plenes de dolçor e de goig, **me van dir** los arquebisbes unes novelles d'on io fui molt dolent e ia molt més trist:

(Per 50) e adoncs los **vaig dir**, tot en plorant:

(Parl)

Anar + Inf

(Parl 69, 8–9) quan Godofrè de Billó **anà conquistar** en la Terra Santa

(EpC)

Anar + Inf

(Ep2 145) Per tal, vostra mare vos prega digau a vostra sogra que li faça una lletra e que no li'n diga cosa niguna, car si ella sabia la bona amor que li aporta a vostra sogra e a sa neboda, no farleu ne diríeu lo que dieu segons diu lo Vallcanera; a bé que nosaltres la **vam pregar** que no li'n diga cosa ninguna, e així ho ha promés, no sé qué es farà.

Anar a + Inf

(Ep2 178) Més, l'avís com lo mateix dia que arribí, **aní a cercar** a mossènyer mossèn Mòger, i trobí'l i parlí ab ell contà'm tota la sua desaventura.

(Ger)

Anar a + **Inf**

(Ger 74) E llavors los jurats y racional y altres officials **anaren** a gran pressa al palau del bisbe **a dir** al que regia per official que donàs lo home al poble per estalviar major mal.

(Ger 84) Los (Ger 85) perayres, sabent que estaven ajustats, **anaren** a dita confraria dels fusters **a dir-los** que per ninguna cosa del món no dexassen de fer son alarde y ordenança, que ells ab los altres officis los defençarien, de manera que ningú los gosaria dir res.

(Ger 112) [32. Combaten casa del virrey y altres maldats] Y encontinent tota la gent y ciutat fonch posada en armes y **anaren** a casa del virrey **a combatre-la** ab atambors y banderes e infinita gent.

(Ger 149) les quals per los de Llíria foren estades derrocades y cremades y per manaments del rey **anaren a fer-les tornar a alsar**. Y arribats al lloch aon estaben, feren-les alsar de nou.

(Ger 151) Y fet açò, **anaren a dinar** a Benisanó.

(Ger 154) Disapte a de dit mes e any, **anaren** los mateixos del dijous al mercat **a derrocar** la barraqueta del carb ahon estava la guarda del carbó, y li manaren que no prengués ningun diner a ningú, sots grans penes.

(Ger 166) Lo sendemà se ajustaran molts dels del poble que tenien per de ses cases, y **anaren** a casa del mestre de Montesa **a supplicar-li** [que] ell volgués pendre treball de anar a la vila de Dénia ab alguns caps del poble, per a fer venir al virrey a València per a que ab sa presència tothom se asosegaria.

(Ger 169) Y de allí **anà a visitar**, lo marquè a dona Anna de la Serdal sa cunyada, y fonch la primera vista que lo dit marquès de Atzeneta se véu ab sa cunyada.

(Ger 175–176) E dos dies aprés que lo dit don Pedro Canoguera fonch en dit castell, partí de Alzira lo camp del agermanats y **anà a posar** siti sobre lo dit castell de Corbera, y començà a dar-li conbat.

(Ger 189) [Agermanats de Valencia enviaren a socórrer als de la Plana y lo duch de Sogorb **anà a favorir** als comanadors]

(Ger 236) Lo duch de Gandia, tantost que fonch desembarcat, partí per la posta per a Castella, y **anà a parlar** ab lo governador general que sa Magestat havia dexat en Castella quant partí per a Alemanya,

(Ger 252) Y aprés de haver feta la volta per tota la ciutat, **anà a descavalcar** al palau de l'arquebisbe ahon estigué tot lo temps que estigué en València.

(Ger 270) com ho féu, que partí de la vila de Nulles, perquè ell allí estava, y, arribat que fonch en Morvedre, **anà a parlar** ab los dits sos amichs, ab los quals tractà que li donassen lo castell.

(Ger 277) Y entraren per lo portal de Serrans y, feta la volta per la ciutat, **anaren a descavalcar** al Real.

(Ger 280–281) **anà a posar** siti sobre Alzira, la qual estava desobedient a l'emperador y al virrey.

(Ger 288) Y la major part de la gent dels que **anaren** a sa casa **a dir-li** que ixqués,

(Ger 291) Y ans de anar a Xàtiva **anaren** primer **a visitar** al virrey y **a tractar** ab ell del negoci, perquè lo virrey per pacte no podia entrar en Xàtiva.

(Ger 298) E arribat que fonch en Valènsia, **anà a posar** al palau de l'archebisbe.

(Ger 301) En aquest endemig, lo traydor de Vicent Pèriz **anà a parlar** ab lo bisbe de Sogorb don Gilabert Martí y li digué que

(Ger 318) Agermanats de València enviaren a socórrer als de la Plana y lo duch de Sogorb **anà a favorir** als comanadors.

(Ger 341) E així com he dit, pasaren a Alcira, de hon **anaren a combatre** lo castell de Corbera perquè era del duch;

(Ger 351) Emperò los de Dénia, encara que **anaren a agermanar-se**, no's pogueren concordar ab lo capità, perquè tots y eren per la roba, que'ls de Dénia tenien molta roba de caballers que allí s'eren embarcats, d'ella furtada, d'ella obligada, d'ella acomanada,

(Ger 357) E de aquí tirà la volta de València e per Moncada **anà a posentar** son real a Paterna, hon féu ajust de huit mília infants e de mil de a caball.

(Ger 388) E lo disapte de matí entrà en València a cavall, e **anà a visitar** la reyna Germana ja dita, que posaba en lo palau del Bisbe.

(Ger 397) E **anà** l'emperador ab tota la flota **a desembarcar** a Gènova,

(Ger 404) E de ací **anà a tenir corts** en Montsó, e duraren les corts fins a festes de Nadal.

Notes

Chapter 1

1. Grammaticalization is traditionally understood as a process through which a lexical item advances from a lexical status to a grammatical one or from a less grammatical status to a more grammatical one.

2. Throughout this book I refer to the constructions under study in the following ways. I use the form *anar* + Inf when referring to the Catalan construction. Although this construction had a prepositional version in the medieval language state (*anar a* + Inf), indication of the preposition would be disturbing, given that the topic of this book is the grammaticalization of the version without a preposition. In contrast, in order to refer to the Spanish construction I will use the form *ir (a)* + Inf. I consider it reasonable to indicate the existence of the preposition, because the grammaticalized construction is used with the preposition *a* in present-day Spanish (*ir a* + Inf). However, I bracket the preposition, since in most of the cases the construction appears without a preposition in the old texts (*ir* + Inf). When I write about the present-day Spanish construction, I do not use the brackets (*ir a* + Inf), since in this language state the version without a preposition is not in use anymore. I indicate the meanings of the Catalan verb *anar* and Spanish verb *ir* when it is reasonable (*anar* 'go' + Inf, *ir (a)* 'go' + Inf). I use the form 'go' + Inf when the only important thing is the meaning of the conjugated verb, and in this way I can refer to the Catalan and Spanish constructions together.

3. The verb *pensar* 'think' appears in the Catalan texts of the corpus mainly with the spelling *pensar*, sporadically as *penssar* and *pessar*, while in Spanish texts it appears written as *pensar* or *penssar*. For reasons of simplification, throughout this book I use the spelling *pensar*; however, when citing examples, I follow the spelling as it appears in the given historical text. The construction formed with the verb *pensar* 'think' can appear with or without a preposition in the medieval language state. For the sake of clarity I use uniformly the form *pensar (de)* + Inf when referring to the construction, both in Catalan and Spanish. When necessary, I also indicate the meaning of *pensar*

(*pensar (de)* 'think of' + Inf). I discuss the use of different versions with or without a preposition in section 2.6.3.3. I use the form 'think of' + Inf when it is the meaning of the conjugated verb which matters, and in this way I can refer to the Catalan and Spanish constructions together.

4. The verb *començar* 'begin' appears in written form as *començar* or *comensar* in the Catalan texts of the corpus, while it appears as *començar, conpeçar* or *compeçar* in the Spanish texts. For sake of simplicity I will use uniformly the spelling *començar*; however, when citing a fragment of a given text, I follow the original spelling. The construction formed with the verb *començar* 'begin' had different versions with or without a preposition in the language state under study, which I discuss in more detail in section 2.6.3.4. Throughout the present book I spell the construction uniformly as *començar (a/de)* + Inf, both in Catalan and Spanish. When it is relevant, I indicate the meaning of the verb *començar* (*començar (a/de)* 'begin to' + Inf). I use the form 'begin to' + Inf when only the meaning of the conjugated verb matters, and in this way I can refer to the Catalan and Spanish constructions together.

5. Four historical texts (Jau, Munt, Per, Cid) in my corpus have an English translation available. In these cases, I used the English version to translate some of the historical text fragments. However, given that they are literary translations, sometimes I have had to modify the English version for the linguistic purposes of the present discussion. For example, Goodenough translates the occurrence *anaren combatre* in Example (69) by the paratactic sequence 'they went and attacked'. I have not adopted this translation because it does not reflect the idea that we are dealing with a periphrastic form. Bibliographic data for translations are given in the 'Historical Sources'.

6. The Catalan Parliament

7. Guilds of artisans in the Kingdom of Valencia, literally: 'brotherhoods'

Chapter 2

1. Similar future periphrases exist in French and Portuguese, too.

2. For further possible typological parallels, see Section 4.3, where a possible evolution of the Catalan construction into a textual marker will be discussed.

3   The paradigm of the Catalan verb *anar* is presented here relying on Juge (2006: 314). As for the forms in brackets, Juge does not indicate in which regions they are in use. Badia i Margarit (1995: 546) notes in his grammar that they pertain to spoken language and are disallowed in normative language use.

4. Pérez Saldanya (1996: 101–102) claims that the modern Catalan *anar* + Inf construction cannot be used with the verb *deure* 'must', but Juge (2002) provides evidence to the contrary.

5. Cf. the 1Pl *anem* and 2Pl *aneu* of the full lexical verb *anar* 'go', the 1Pl *vem*

and 2Pl *veu* and the *var-* forms of the auxiliary. These latter were created by analogy, modelled on regular preterit forms of first conjugation verbs (cf. Juge 2006).

6. Except in the Catalan dialect of the Baleares, where the preposition *a* does not appear even in this meaning (Segura-Llopes 2012: 119 n. 3).

7. That is, it has the same meaning as *anar* + Inf had originally.

8. In order to illustrate modal uses of the construction, let us consider Bravo Martín's (2008) examples. Epistemic modality: *¡{Qué/Cómo} va a conocerlo!* 'No lo conoce' (Bravo Martín 2008: 221). Literally: 'He **is not going to know** it/him!' = 'He does not know it/him.' Deontic modality: *Los invitados no van a comer calamares y los anfitriones salmón.* 'Los invitados **no pueden comer** calamares y los anfitriones salmón' (Bravo Martín 2008: 230). Literally: 'The guests **are not going to eat** calamary and the hosts salmon' = 'The guests **cannot eat** calamary and the hosts salmon'.

9. Cf. 'Peter goes to the beach in order to learn'. For arguments of a verb included in the conceptual-semantic representation, but lexically unrealized, see Bibok and Németh T. (2001a).

10. The sources of the historical examples are abbreviated at the end of each example, with page and/or line numbers (full details can be found in Historical Sources).

11. The symbol '=' is used to indicate literal meaning, while the symbol '?' indicates strange/nonsensical readings.

12 *La chanson de Roland.* Paris: Larousse, 1972.

13 I thank Csilla Ilona Dér who suggested I consider this theoretical possibility.

14. On demographic movement from Occitania to Catalonia in the Early Modern Period, see Jacobs and Kunert (2014: 198).

15. The influence of Gascon, a variety of Occitan.

16. It is not intended to mean here that Catalan would have acquired the fully grammaticalized construction as a past tense. This assumption is contradicted by historical data which show the use of the construction in highly limited contexts in the earliest texts. The strongest supposition I can accept is that Catalan adopted some conventionalized local schemas (e.g. 'go to hit/attack'), which were used to express some kind of aspectual or intentional meaning in certain contexts.

17. As I have mentioned above, Berchem (1968) and Jacobs and Kunert (2014) describe that the periphrasis is still in use by a Franco-Provençal colony in Guardia Piemontese (Italy). Berchem (1968) argues that this variety acquired the already grammaticalized construction, while Jacobs and Kunert (2014) regard the presence of the *go*-past in this variety as an archaic feature, which traces back to Old Occitan.

18. A criticism of this observation is provided in Mendeloff (1968). However, he does not offer any alternative hypothesis and considers the phenomenon as a mystery waiting to be solved.
19. A more detailed variant of the same hypothesis is provided in Pérez Saldanya and Hualde (2003).
20. These forms actually do not only express tense, but also aspect and mood. Juge (2006) adopts the term *tamcat* to refer to a set of forms with a given tense/aspect/mood combination.
21. 'buidat del seu contingut semàntic concret de verb de moviment, *anar* té com a funció única, pel seu caràcter dinàmic, animar el relat i no pas marcar cronològicament el començament o l'acabament d'una acció'.
22. Consider, for instance, the following: the French *se mettre à faire qc.* 'start doing sg.' < 'sit down to do sg.'; the Italian *mettersi a fare qc.* 'start doing sg.' < 'sit down to do sg.'; the Spanish *ponerse a hacer algo* 'start doing sg.' < 'put oneself in a position to do sg.'; the English *start doing sg.* < Middle English *sterten* 'jump, leap'; the Latin *incipere* 'begin' and the German *anfangen* 'begin' < 'catch, grasp' (Detges 2004: 214).
23. As for the grammaticalization paths described by Bourdin, see Section 4.3; here I discuss only aspects concerning deixis.
24. Theoretical considerations concerning reanalysis are presented below, in Section 3.6.3.3.
25. For the reason why this framework is called a theory, see Heine (2003/2005: 575, 577).
26. I thank Enikő Németh T. for this interesting observation (personal communication).
27. As Kearns (2002: 2) notes, this is not general: works on the role of implicatures in semantic change usually do not rely directly on Grice, but on one of the neo-Gricean theories, e.g. Horn (1984) or Atlas and Levinson (1981).
28. In Section 1.2 I have introduced my object-scientific research questions in general, here I present them in more detail.
29. Aragonese and Castilian are geographically connected to the Western Catalan dialects, which also include Valencian.
30. Occurrences used in present reference contexts are not included in the research, since they do not figure in the tables.
31. Segura-Llopes (2012: 131) provides similar results concerning the spread of present tense variants.
32. According to Csala (2002: 111), Christian greeting etiquette allowed the kiss on the lips, but among Muslims only a kiss on the shoulder, neck, or upper arm was permitted.
33. The finite verbs of the periphrases mostly occur in the perfective past in

the contexts discussed here. However, the periphrases formed with 'go' and 'think' have some occurrences which show the finite verb in the imperfective past. Occurrences of the THINK OF-periphrasis with the finite verb in the imperfective seem to refer to a performed action, similarly to its perfective past occurrences. In contrast, as Pérez Saldanya (1996: 89) points out, the verb 'go' conjugated in the imperfective past suggests that the subject did not manage to perform the act described in the infinitive. The verb 'go' in the GO-periphrasis, when conjugated in the imperfective past, seems to retain its original motion meaning in narrative texts, i.e. these contexts are not relevant in the present discussion. The verb 'begin' in the BEGIN TO-periphrasis does not occur in the imperfective past in the present corpus.

34. The following Spanish constructions represent these types: 1. the present perfect: 'the present indicative of the verb *haber* + past participle', 2. the construction *volver a* + Inf (aspectual meaning, 'to do something again') expresses repetition, the construction *deber de* + Inf (modal meaning) indicates probability or possibility, 3. the construction *tardar en* + Inf 'to take a long time to do something'.

35. The author probably uses the term *desemanticized* in the sense that the verbal form is used not independently, in its original lexical meaning, but the whole construction expresses a unified meaning.

36. Cf. contexts of grammaticalization of '*anar* + Inf' and Examples (73)–(81), in Section 3.4.2.2.

37. Colon mentions the following infinitives among others: *eixir* 'to go out', *gitar* 'to turn around', *abraçar* 'to embrace', *besar* 'to kiss', *ferir* 'to attack/hit/wound' and *trencar* 'to break'.

38. *Grammaticization* is Bybee's term for grammaticalization. For different denominations of grammaticalization, see Heine (2003/2005: 577).

39. For the notion of frequency, cf. Section 3.6.3.2. In running text, I use the most frequent spelling form of the infinitives, but when quoting historical occurrences, I write them as they appear in the given location in the corpus.

40. There are prepositional and non-prepositional uses of the constructions formed with the verbs 'go', 'think', and 'begin', both in medieval Catalan and Spanish (for details, see Sections 2.5.2, 2.6.3.3, and 2.6.3.4). In this section, dedicated to entrenched schemas, I follow the method of referring to the constructions as presented in Chapter 1 n. 2. For instance, in spite of the fact that in Spanish texts the preposition *a* does not appear in the combination of the verb *ir* with the infinitive *ferir*, I refer to this construction as *ir (a) ferir*, because occurrences of the prepositional version *ir a ferir* are assumed to be found in a larger corpus (cf. also considerations presented in Chapter 1 n. 2). I refer in a similar way to the constructions formed with the verbs 'think'

and 'begin'. Namely, in spite of the fact that all occurrences of *pensar (de) ferir* contain the preposition *de* in my corpus, I use the parenthetical form when referring to the combination of the verb *pensar* with the infinitive *ferir* (cf. Section 2.6.4.2), since a larger corpus may contain occurrences without a preposition as well (as suggested by the non-prepositional combinations of *pensar* with other infinitives).

41. A small shield.

42. There is an example with the spelling *pendra*.

43. There is an occurrence spelt *abrassar*.

44. The form *.I.* is the way of writing *un* 'a/an (indefinite article)/one' used in some historical texts.

45. There are different spellings of this infinitive: *assetiar, assetgar, asetgar, asetyar*. The three occurrences of the phrase *anar tenir setge* 'go to besiege' also belong here, since convey an equivalent meaning.

46. Since there are different spellings of city names in the historical texts of the corpus and many cities have different names in Catalan and Spanish, I always use the spelling of the original text in the translations, unless there is a widely accepted English name of the city.

47. Sometimes the variant *veure* can also occur.

48. Sometimes the variant *veure's* can also occur.

49. Or perhaps in all the cases when the finite verb and the infinitival complement are separated from each other. Note that the preposition only appears before the first infinitive. When two or more infinitives are coordinated, the preposition seems to be obligatory only before the first one, as confirmed by other examples as well.

50. It also appears with the following spellings: *uer, ueer*, and *veyer*.

51. In present-day Spanish the periphrasis is used without a preposition in an intentional sense, with the preposition *en* when the literal meaning 'think of' is intended, and with the preposition *de* when the construction expresses opinion. The intentional periphrasis can be used in current Catalan without a preposition or with the preposition *de*.

52. The preposition *a* still continues to be used in the current language state both in Catalan and Spanish.

53. The notion of *broad context* is discussed in more detail in Section 3.4.2.3.

54. Colon (1976/1978: 141) mentions a similar use in medieval French as well, where the collocation *va dire* introduces the direct quotation of the content of communication. However, he also quotes an occurrence *va commencer à dire* '(she/he) began to say' (= '(she/he) goes to begin to say'), which testifies that the GO-construction is not enough in itself to produce the intended effect.

55. At this stage, principally telic verbs appear in the construction. However, Segua-Llopes (2012: 35) emphasizes that their use is not exclusive, there are also some examples with activity verbs.
56. Bruguera (1981: 41) admits three of them.
57. For linguistic clues excluding the source meaning, see below in Sections 3.5.2 and 3.5.3.
58. Cf. the discussion of Example (31), as well.
59. The *Companyia Catalana* (the *Catalan Company*) was a free company of mercenaries founded by Roger de Flor in the early 14th century.

Chapter 3

1. These research questions are the more specific expositions of my general methodological questions summarized in Section 1.2.
2. The traditional use of the term *data* raises several theoretical and methodological problems, which I discuss below in Section 3.6.2.1. Until then I use an intuitive notion of data, as appears in the literature referred to.
3. A similar discussion of data types in pragmatics is provided in Németh T. (2006).
4. Fortunately, we have some kind of metalinguistic evidence – although its extent is limited – concerning the use of the Catalan *anar* + Inf construction, as well.
5. Although Fischer's suggestion is very useful, we must take into account all the problems which the use of quantitative information evokes, for instance, that the frequency of occurrence cannot be used as a criterion for grammaticality.
6. '[Cal evitar l'ús de] *Vaig anar* e *vaig venir* per *aní* e *venguí* e semblants.'
7. 'Regla 48: aquests vocables de *vaig anar* a misser Hierony Pau ne a mi, Pere Miquel Carbonell, no par sian bons vocables. Més val dir: *anam, venguem*; no: *vam anar.*'
8. 'That person told me these words', where the meaning 'told' is expressed by the analytic past *va dir*.
9. 'That person told me these words', where the meaning 'told' is expressed by the synthetic past *dix*.
10. 'La segona manera és com algú posa en sos dictatz, per manera recitatòria las paraulas següens: "Aquelh *me va dir* aytals paraulas." Con bastaria e seria pas belh e sens pedaç que dixés: "Aquelh *me dix* aytals paraulas", car si bé guardatz, aytal sentència han aquestas paraulas darreras com las primeras, e ab aytant aquelhas duas diccions, ço són, *me va*, romanen del tot pedaç, per tal com són supèrfluas.' *Torcimany*, I, p. 168, § 33.
11. '*Il luy va dire* se met vulgairement pour *il luy dit.*'
12. Cf. metalinguistic evidence provided in Colon (1976/1978: 147–148) con-

cerning the corresponding Provençal construction, presented as similarly stigmatized.

13. Heine takes this term from Evans and Wilkins (1998: 5).

14. In the presentation of Diewald's (2002) typology of contexts, I use the term *conversational implicature* following her terminology. A more detailed discussion of the use of the terms *pragmatic inference* and *conversational implicature* has been provided in Section 2.4.2.

15. The use of plural here is a majestic plural, as well as in Examples (79), (84), and (85).

16. The *fonevol* was a type of war-engine, which threw very large stones. The *brigola* also was an engine designed to throw stones. James I mentions other similar machines in his chronicle, as well (e.g. *almangenech, manganel, trabuquet*) (see Kagay 2010: 93).

17. In the present investigation, this finite verb is the finite verb form of the lexeme under study (GO). In other studies, another conception of narrow context might be necessary, which also allows for contexts which do not contain a finite verb.

18. Analyses of occurrences presented in Chapter 2 also contained linguistic clues which excluded the source meaning; e.g. the analyses of occurrences in (65) and (72).

19. One of the anonymous reviewers of my article (Nagy C. 2010b) which also included the example in (82), commented that it (and also Example (91)) can be considered as an instance of bridging context in the sense that although the verb *anar* does not retain the original meaning of displacement from one place to another, it could indicate the movement of part(s) of the body. This is an interesting point; however, these examples should rather be considered as instances of switch context, because in the case of bridging context the old lexical meaning can also be assigned, which is not true of these occurrences.

20. *Almogavar* is the name of a class of soldiers from the Kingdom of Aragon (and other Iberian kingdoms) during the 13th and 14th centuries.

21. A kind of sword, from Bordeaux

22. Colon (1959/1978: 124–125) provides a similar example from Desclot's chronicle, where it is the preceding clause that contains the verbal form in the past perfect. Also, Colon concludes that the periphrasis has a past tense value in this context.

23. Unless otherwise noted, verbal forms are in the indicative.

24. In this section I discuss the use of corpus data in historical linguistics in general. For specific difficulties in historical pragmatic research, see Sections 3.4 and 3.5.

25. On the notion of evidence, see also Kertész and Rákosi (2008).

26. Even in these cases it is difficult (if not impossible) to judge whether the content of change of location is present as an inferential content, or forms part of the coded meaning of the periphrasis.
27. For constructions in grammaticalization, see Traugott (2003) and Bybee (2003/2005: 602–603).
28. However, reanalysis and category conversion should be differentiated. Among others, Lehmann (2005: 11) also makes a distinction between *lateral conversion* and category changes in grammaticalization.
29. For different approaches to the relationship between grammaticalization and reanalysis, see Dér (2008: 103–105).
30. The consideration of every hypothesis and author mentioned in this volume is not possible now and probably would not produce new results.
31. The Catalan version is as follows: 'Lo castor sí és una bístia qui hi ha un membre lo qual és de gran virtut, los quals membres són los seus genitius. E con aquest castor és cassat per los cassadors, e·ls cans lo aconseguexen, ne coneix la rahó per què ell es cassat, **va's pendre** los genitius ab les dents, e arranque'ls-se, e gita'ls en terra. E·l cassador pren los genitius per què ell lo cassa, e lexa anar lo castor.' 'The beaver, verily, is an animal that has a member which is very precious, and it is its testicles. And when pursued by the hunters and the dogs are about to catch it up, the beaver understands the reason why it is hunted and **bites (= goes to take)** the testicles with its teeth, chews them off and throws them to the ground. The hunter takes the genitals which he hunts it for, and lets the beaver go.' The expression *va's pendre* 'takes' (= ?'goes to take') in the Catalan translation corresponds in the original Provençal text to the verbal form *pren* 'takes', which is a simple present verbal form. The use of the Catalan construction *anar* + Inf is supposedly a way to make the description more animated, emphasizing the quick and unexpected nature of the act described. Colon (1959/1978: 128) argues that the Catalan translation is more expressive than the original version.
32. Colon (1959/1978: 130) himself notes that before 1350 only a certain type of infinitive could enter the periphrasis, which shows that the construction was not yet completely grammaticalized.
33. These examples are provided at another point of his argumentation and with a different aim from that of the present discussion.
34. The segment from Muntaner's Chronicle is the following: 'E tantost que açò hac fet, el cavall se sentí ferit, e llevà's davant e detràs aixi que fóra caüt si no fos que era ab cadena fermat en la sella. Què us diré? Ella **va metre** la man a l'espaa, e venc a un portell altre, e **anà ferir** lo cavall per la testera; e el cavall estec estabornit. Què us diré? Lo cavall **va pendre** per la regna e cridà: "Cavaller, mort sóts, si no us retrets." E el cavaller tenc-se per mort, e **va tenir**

lo bordó e **va'l gitar** en terra, e <u>retré's</u> a ella; et ella <u>pres</u> lo bordó e <u>trasc</u>-li la llança de la cuixa, e així <u>mès</u>-lo dins Peralada.' 'And as soon as she <u>had done</u> this and the horse <u>felt</u> himself wounded, it <u>bucked</u> and <u>reared</u>, so that he [the French knight] <u>would have fallen</u>, if he <u>had not been chained</u> to the saddle. What <u>shall</u> I <u>tell</u> you? She **took (= ?goes to take)** hold of her sword and <u>aimed</u> at another opening and **wounded (= went to wound)** the horse in the head; and the horse <u>was stunned</u>. What <u>shall</u> I <u>tell</u> you? She **seized (= goes to seize)** the horse by the reins and <u>cried</u>: 'Knight, you <u>are</u> a dead man if you <u>do not surrender</u>.' And the knight <u>thought</u> himself a dead man, he **took (= ?goes to take)** the bordon and **threw (= ?goes to throw)** it down and <u>surrendered</u> to her; and she <u>took</u> the bordon, and then <u>pulled</u> out the lance from his thigh and so she <u>brought</u> him in to Peralada.'

Chapter 4

1   For a detailed criticism of this position, see Juge (2006: 327–333).
2. Cf. the following Spanish example: *El petardo fue a explotar al lado de mi ventana.* '(Finally) The firecracker exploded (right) beside my window.' In this context the construction conveys a counter-consecutive meaning as described in Bourdin (2008), inasmuch as it describes the event as the most unexpected option.
3. Bourdin (2008) uses the following abbreviations in the glosses: 1 – first person, 3 – third person, SG – singular, PL – plural, PRF – perfect, M – masculine, OBJ – object, PST – past.

Historical sources

Catalan texts

(Desc) Bernat Desclot, *Crònica* [Chronicle] (Vols. 1–5). Barcelona: Edicions 62, 1990.

(EpC) Max Cahner (ed.), *Epistolari del Renaixement* [Collection of Renaissance Letters]. Vol. 1: 62–133, 137–198; Vol. 2: 76–77, 88–126.

(Ger) Eulàlia Duran (ed.), *Cròniques de les germanies: Les cròniques valencianes sobre les germanies de Guillem Ramon Català i de Miquel Garcia* [History of the Germanies: History of the *germanies* of Valencia of Guillem Ramon Català and Miquel Garcia]. València: Eliseu Climent.

(Jau) *Llibre dels fets del rei en Jaume* [The Book of Deeds of James I of Aragon] (Vols. 1–2). Barcelona: Barcino, 1991. English translation: *James I (The Conqueror) King of Aragon: Chronicle* (trans. Thomas Forster). Cambridge, Ontario: In Parentheses Publications, 2000. Available at: http://www.yorku.ca/inpar/jaume_forster.pdf.

(Munt) Ramon Muntaner, *Crònica* [Chronicle] (Vols. 1–2). Barcelona: Edicions 62, 1979/1990–1991. English translation: Ramon Muntaner, *Chronicle* (trans. Lady Goodenough). Cambridge, Ontario: In Parentheses Publications, 2000. Available at: http://www.yorku.ca/inpar/muntaner_goodenough.pdf.

(Parl) Ricard Albert and Joan Gassiot (eds.), *Parlaments a les corts catalanes* [Speeches at the Catalan Courts]. Barcelona: Barcino, 1928.

(Per) Ramon de Perellós, Viatge del vescomte Ramon de Perellós i de Roda fet al Purgatori nomenat de Sant Patrici [The Journey of Viscount Ramon de Perellós i de Roda to Saint Patrick's Purgatory]. In *Novel·les amoroses i morals*. Barcelona: Edicions 62, 1982. English translation: *The Journey of Viscount Ramon de Perellós to Saint Patrick's Purgatory* (trans. Alan Mac an Bhaird). Cork: CELT, 2012. Available at: http://www.ucc.ie/celt/published/T100079A/.

Spanish texts

(Alf) *Crónica de Alfonso X* [Chronicle of Alfonso X]. In ADMYTE (Archivo Digital de Manuscritos y Textos Españoles), CNUM 302.

(Arn) *Tratado de amores de Arnalte y Lucenda* [Treatise Concerning the Loves of Arnalte and Lucenda]. In ADMYTE (Archivo Digital de Manuscritos y Textos Españoles), CNUM 2995.

(Cid) Unknown author, *Cantar de mio Cid* [The Poem of the Cid]. Madrid: Espasa Calpe, 1976. English translation available at: http://miocid.wlu.edu/.

(CrPC) *Crónica popular del Cid* [Popular Chronicle of El Cid]. In ADMYTE (Archivo Digital de Manuscritos y Textos Españoles), CNUM 6993.

(GrCo) *Gran conquista de Ultramar* [The Great Conquest of Outremer]. In ADMYTE (Archivo Digital de Manuscritos y Textos Españoles), CNUM 484.

(Mar) *Vida de Santa María Egipciaca* [The life of Saint Mary of Egypt]. In ADMYTE (Archivo Digital de Manuscritos y Textos Españoles), CNUM 628.

(Pier) *Pierres y Magalona: La historia de la linda Magalona y del muy esforzado caballero Pierres de Provenza* [Pierres and Magalona: The story of the beautiful Magalona and the very brave knight Pierres of Provance]. In ADMYTE (Archivo Digital de Manuscritos y Textos Españoles), CNUM 6280.

(Tres) *Libre dels tres reys d'Orient* [Book of the three Kings of the Orient]. In ADMYTE (Archivo Digital de Manuscritos y Textos Españoles), CNUM 527.

(Tris) *Cuento de Tristán de Leonis* [The story of Tristan of Leonis]. In ADMYTE (Archivo Digital de Manuscritos y Textos Españoles), CNUM 389.

References

Alonso, Amado (1982) Sobre métodos: construcciones con verbos de movimiento en español. In Amado Alonso, *Estudios lingüísticos: Temas españoles* 190–236. Madrid: Gredos.

Atlas, Jay David and Levinson, Stephen C. (1981) *It*-clefts, informativeness, and logical form: Radical pragmatics (Revised Standard Version). In Peter Cole (ed.) *Radical pragmatics* 1–61. New York: Academic Press.

Badia i Margarit, Antoni M. (1950) 'Regles de esquivar vocables o mots grossers o pagesívols.' Unas normas del siglo XV sobre pureza de la lengua catalana. *Boletín de la Academia de Buenas Letras de Barcelona* 23: 137–161.

Badia i Margarit, Antoni M. (1981) *Gramàtica històrica catalana*. València: Tres i Quatre.

Badia i Margarit, Antoni M. (1995) *Gramàtica de la llengua catalana: Descriptiva, normativa, diatòpica, diastràtica*. Barcelona: Edicions Proa.

Bauhr, Gerard (1989) *El futuro en –RÉ e 'IR' A + infinitivo en español peninsular moderno*. Goterna: Kungälv.

Berchem, Theodor (1968) Considérations sur le parfait périphrastique *vado + infinitif* en catalan et gallo-roman. In Antonio Quilis Morales, Ramón B. Carril and Margarita Cantarero (eds.) *Actas del XI Congreso Internacional Lingüística y Filología Románicas* (Vol. 3) 1159–1170. Madrid: CSIC.

Biber, Douglas and Finegan, Edward (eds.) (1994) *Sociolinguistic Perspectives on Register*. Oxford: Oxford University Press.

Bibok, Károly (2004) Word meaning and lexical pragmatics. *Acta Linguistica Hungarica* 51(3–4): 265–308. https://doi.org/10.1556/ALing.51.2004.3-4.3

Bibok, Károly and Németh T., Enikő (2001a) How the lexicon and context interact in the meaning construction of utterances. In Enikő Németh T. and Károly Bibok (eds.) *Pragmatics and the Flexibility of Word Meaning* (Current Research in the Semantics/Pragmatics Interface 8) 289–320. Amsterdam: Elsevier.

Bibok, Károly and Németh T., Enikő (2001b) On the interaction between lexical and contextual information in the composition of utterance meaning. In Enikő Németh T. (ed.) *Cognition in Language Use: Selected Papers from the 7th International Pragmatics Conference* 12–25. Antwerp: International Pragmatics Association.

Bourdin, Philippe (2008) On the grammaticalization of 'come' and 'go' into markers of textual connectivity. In María José López-Couso and Elena Seoane (eds.) *Rethinking Grammaticalization: New Perspectives* 37–59. Amsterdam: John Benjamins. https://doi.org/10.1075/tsl.76.05bou

Bravo Martín, Ana (2008) *La perífrasis 'ir a + infinitivo' en el sistema temporal y aspectual del español.* PhD Thesis, Madrid. http://eprints.ucm.es/8074/1/T30424.pdf.

Bres, Jacques and Labeau, Emmanuelle (2013) The narrative construction *va + infinitive* in contemporary French: A linguistic phoenix rising from its medieval ashes? *Diachronica* 30(3): 295–322. https://doi.org/10.1075/dia.30.3.01bre

Bruguera, Jordi (1981) La locució prepositiva 'de part', el present històric i el perfet perifràstic en la Crònica de Jaume I. In *Estudis de llengua i literatura catalanes: Vol. 3. Miscel·lània Pere Bohigas I* 27–42. Barcelona: Publicacions de l'Abadia de Montserrat.

Bybee, Joan L. (2002) Cognitive processes in grammaticalisation. In Michael Tomasello (ed.) *The New Psychology of Language* (Vol. 2) 145–167. Hillsdale, NJ: Lawrence Erlbaum.

Bybee, Joan L. (2003/2005) Mechanisms of change in grammaticization: The role of frequency. In Brian D. Joseph and Richard D. Janda (eds.) *The Handbook of Historical Linguistics* 602–623. Oxford: Blackwell.

Bybee, Joan L. (2005) From usage to grammar: The mind's response to repetition. LSA Presidential Address, 8 January 2005. http://www2.sfs.uni-tuebingen.de/~gerhard/lehre/ss08/exemplarBased/Bybee05.pdf.

Bybee, Joan (2007) Diachronic linguistics. In Dirk Geeraerts and Hubert Cuyckens (eds.) *The Oxford Handbook of Cognitive Linguistics* 945–987. Oxford: Oxford University Press.

Bybee, Joan L. and Pagliuca, William (1987) The evolution of future meaning. In Anna Giacalone Ramat, Onofrio Carruba, and Giuliano Bernini (eds.) *Papers from the 7th International Conference on Historical Linguistics* 109–122. Amsterdam: John Benjamins. https://doi.org/10.1075/cilt.48.09byb

Bybee, Joan L., Perkins, Revere, and Pagliuca,William (1994) *The Evolution of Grammar: Tense, Aspect and Modality in the Languages of the World.* Chicago: The University of Chicago Press.

Bynon, Theodora (1983) *Historical Linguistics.* Cambridge: Cambridge University Press.

Calvo Pérez, Julio (1995) El TAMP en valencià. *Caplletra* 19: 259–278.

Colon, Germà (1959/1978) El perfet perifràstic català 'va + infinitiu'. In Germà Colon (ed.) *La llengua catalana en els seus textos* (Vol. 2) 119–130. Barcelona: Curial.

Colon, Germà (1976/1978) Sobre el perfet perifràstic 'vado + infinitiu', en català, en provençal i en francès. In Germà Colon (ed.) *La llengua catalana en els seus textos* (Vol. 2) 131–174. Barcelona: Curial.

Coromines, Joan (1980–1991) *Diccionari etimològic i complementari de la llengua catalana.* Barcelona: Curial Edicions Catalanes.

Csala, Károly (trans.) (2002) *Ének Çidről. Középkori spanyol epikus költemény.* [Hungarian translation of *Cantar de mio Cid*] Budapest: Eötvös József Könyvkiadó.

De Vogelaer, Gunther (2010) Does grammaticalization need analogy? Different pathways on the 'pronoun/agreement marker' cline. In Katerina Stathi, Elke Gehweiler, and Ekkehard König (eds.) *Grammaticalisation: Current Views and Issues* (Studies in Language Companion Series 119) 221–240. Amsterdam: John Benjamins.

Dér, Csilla Ilona (2004) Határok nélkül: A grammatikalizáció státusáról [Across the boundaries on the status of grammaticalisation]. *Nyelvtudományi Közlemények* 101: 182–194.

Dér, Csilla Ilona (2005) *Grammatikalizációs folyamatok a magyar nyelvben – Elméleti kérdések és esettanulmányok* [Processes of grammaticalization in Hungarian – Theoretical questions and case studies] PhD Thesis, Budapest, ELTE.

Dér, Csilla Ilona (2008) *Grammatikalizáció* [Grammaticalization] (Nyelvtudományi Értekezések 158). Budapest: Akadémiai Kiadó.

Dér, Csilla Ilona (2013) Grammaticalization: A specific type of semantic, categorial, and prosodic change. *Berliner Beiträge zur Hungarologie* 18: 160–179.

Detges, Ulrich (2004) How cognitive is grammaticalisation? The history of the Catalan *perfet perifràstic*. In Olga Fischer, Muriel Norde, and Harry Perridon (eds.) *Up and Down the Cline: The Nature of Grammaticalisation* 211–227. Amsterdam: John Benjamins.

Diewald, Gabriele (2002) A model for relevant types of contexts in grammaticalization. In Ilse Wischer and Gabriele Diewald (eds.) *New Reflections on Grammaticalization* 103–120. Amsterdam: John Benjamins. https://doi.org/10.1075/tsl.49.09die

Dömötör, Adrienne (2012) A nyelvtörténeti adat: Elvek, gyakorlat, lehetőségek. [Data in historical linguistics: Principles, practice, and possibilities] *Magyar Nyelv* 108(1): 39–51.

Evans, Nicholas and Wilkins, David (1998) The knowing ear: An Australian test of universal claims about the semantic structure of sensory verbs and their extension into the domain of cognition. *Arbeitspapier* 32., NF. Cologne, Institut für Sprachwissenschaft.

Feghali, Michel (1928) *Syntaxe des parlers arabes actuels du Liban* (Bibliothèque de l'École des langues orientales vivantes 9). Paris: Librairie orientaliste Paul Geuthner.

Fischer, Olga (2004) What counts as evidence in historical linguistics? *Studies in Language* 28(3): 710–740. https://doi.org/10.1075/sl.28.3.21fis

Fischer, Olga (2007) *Morphosyntactic Change: Functional and Formal Perspectives.* Oxford: Oxford University Press.

Fischer, Olga (2010) An analogical approach to grammaticalization. In Katerina Stathi, Elke Gehweiler, and Ekkehard König (eds.) *Grammaticalization: Current Views and Issues* 181–219. Amsterdam: John Benjamins. https://doi.org/10.1075/slcs.119.11fis

Fitzmaurice, Susan and Taavitsainen, Irma (2007) Historical pragmatics: What it is and how to do it. In Susan M. Fitzmaurice and Irma Taavitsainen (eds.) *Methods in Historical Pragmatics* (Topics in English Linguistics 52) 11–36. Berlin: Mouton de Gruyter.

Fludernik, Monika (1992) The historical present tense in English literature. An oral pattern and its literary adaptation. *Language and Literature* 17: 77–107.

Forgács, Tamás (1993–1994) Zárt korpuszok és pótkompetencia [Closed corpora and substitute competence]. *Néprajz és Nyelvtudomány* 35: 17–23.

Fortson, Benjamin W. (2003) An approach to semantic change. In Brian D. Joseph and Richard D. Janda (eds.) *The Handbook of Historical Linguistics* 648–666. Oxford: Blackwell. https://doi.org/10.1002/9780470756393.ch21

Gili Gaya, Samuel (1961) *Curso superior de sintaxis española*. Barcelona: Biblograf.

Gómez Torrego, Leonardo (1988) *Perífrasis verbales: Sintaxis, semántica y estilística*. Madrid: Arco/Libros.

Gómez Torrego, Leonardo (1999/2000) Los verbos auxiliares. Las perífrasis verbales de infinitivo. In Ignacio Bosque Muñoz and Violeta Demonte Barreto (eds.) *Gramática descriptiva de la lengua Española* (Vol. 2) 3323–3389. Madrid: Espasa Calpe.

Gougenheim, Georges (1929/1971) *Étude sur les périphrases verbales de la langue française*. Paris: Nizet.

Greule, Albrecht (1982) *Valenz, Satz und Text: Syntaktische Untersuchungen zum Evangelienbuch Otfrieds von Weißenburg auf der Grundlage des Codex Vindobonensis*. Munich: Fink.

Grice, Paul (1975/1989) Logic and conversation. In Paul Grice *Studies in the Way of Words* 22–40. Cambridge, MA: Harvard University Press.

Grice, Paul (1978/1989) Further notes on logic and conversation. In Paul Grice, *Studies in the Way of Words* 41–57. Cambridge, MA: Harvard University Press.

Harris, Martin (1982) The 'past simple' and the 'present perfect' in Romance. In Vincent Nigel and Martin Harris (eds.) *Studies in the Romance Verb* 42–70. London: Croom Helm.

Haspelmath, Martin (1999) Optimality and diachronic adaptation. *Zeitschrift für Sprachwissenschaft* 18(2): 180–205. https://doi.org/10.1515/zfsw.1999.18.2.180

Heine, Bernd (2002) On the role of context in grammaticalization. In Ilse Wischer and Gabriele Diewald (eds.) *New Reflections on Grammaticalization* 83–102. Amsterdam: John Benjamins. https://doi.org/10.1075/tsl.49.08hei

Heine, Bernd (2003/2005) Grammaticalization. In Brian D. Joseph and Richard D. Janda (eds.) *The Handbook of Historical Linguistics* 575–601. Oxford: Blackwell.

Heine, Bernd and Kuteva, Tania (2002) *World Lexicon of Grammaticalization*. Cambridge: Cambridge University Press. https://doi.org/10.1017/CBO9780511613463

Heine, Bernd, Claudi, Ulrike, and Hünnemeyer, Friederike (1991) *Grammaticalization: A Conceptual Framework*. Chicago: University of Chicago Press.

Hock, Hans Henrich (1991) *Principles of Historical Linguistics*. Berlin: Mouton de Gruyter. https://doi.org/10.1515/9783110219135

Hock, Hans Henrich (2003/2005) Analogical change. In Brian D. Joseph and Richard D. Janda (eds.) *The Handbook of Historical Linguistics* 441–460. Oxford: Blackwell. https://doi.org/10.1002/9780470756393.ch11

Holmer, Nils M. (1946) Outline of Cuna grammar. *International Journal of American Linguistics* 12: 185–197. https://doi.org/10.1086/463913

Hopper, Paul J. and Traugott, Elizabeth C. (1993) *Grammaticalization*. Cambridge: Cambridge University Press.

Horn, Laurence R. (1984) Toward a new taxonomy for pragmatic inference: Q-based and R-based implicature. In Deborah Schiffrin (ed.) *Meaning, Form and Use in Context: Linguistic Applications (Georgetown University Round Table on Languages and Linguistics 1984)* 11–42. Washington, DC: Georgetown University Press.

Jacobs, Andreas and Jucker, Andreas H. (1995) The historical perspective in pragmatics. In Andreas H. Jucker (ed.) *Historical Pragmatics* 3–27. Amsterdam: John Benjamins. https://doi.org/10.1075/pbns.35.04jac

Jacobs, Bart and Kunert, Hans Peter (2014) Whatever happened to the Occitan *go-past*? Insights from the dialects of Gascony and Guardia Piemontese. *Revue Romane* 49(2): 177–203. https://doi.org/10.1075/rro.49.2.01jac

Jeffers, Robert J. and Lehiste, Ilse (1979) *Principles and Methods for Historical Linguistics.* Cambridge, MA: MIT Press.

Jucker, Andreas H. (1994) The feasibility of historical pragmatics. *Journal of Pragmatics* 22(5): 533–535. https://doi.org/10.1016/0378-2166(94)90083-3

Jucker, Andreas H. (2009) Speech act research between armchair, field and laboratory: The case of compliments. *Journal of Pragmatics* 41(8): 1611–1635. https://doi.org/10.1016/j.pragma.2009.02.004

Juge, Matthew L. (1999) On the rise of suppletion in verbal paradigm. In Steve S. Chang, Lily Liaw, and Josef Ruppenhofer (eds.) *Proceedings of the 25th Annual Meeting of the Berkeley Linguistics Society* 183–194. Berkeley, CA: Berkeley Linguistics Society. https://doi.org/10.3765/bls.v25i1.1195

Juge, Matthew L. (2002) *Tense and Aspect in Periphrastic Pasts: Evidence from Iberian Romance.* PhD Thesis, University of California-Berkeley.

Juge, Matthew L. (2006) Morphological factors in the grammaticalisation of the Catalan 'go' past. *Diachronica* 23(2): 313–339. https://doi.org/10.1075/dia.23.2.05jug

Juge, Matthew L. (2008) Narrative and the Catalan *go-past*. *Folia Linguistica Historica* 29: 27–56. https://doi.org/10.1515/FLIH.2008.27

Kagay, Donald J. (2010) Jaime I of Aragon: Child and master of the Spanish reconquest. *Journal of Medieval Military History* 8: 69–108.

Kearns, Kate (2002) Implicature and language change. In Jef Verschueren, Jan-Ola Östman, Jan Blommaert, and Chris Bulcaen (eds.) *Handbook of Pragmatics*. Amsterdam: John Benjamins. https://doi.org/10.1075/hop.6.imp1

Kenesei, István (2000) Van-e segédige a magyarban? Esettanulmány a grammatikai kategória és a vonzat fogalmáról. [Is there a class of auxiliaries in Hungarian? A case study on grammatical category and the notion of complement] In István Kenesei (ed.) *Igei vonzatszerkezet a magyarban* [Argument structure in Hungarian] 157–195. Budapest: Osiris.

Kepser, Stefan and Reis, Marga (eds.) (2005) *Linguistic Evidence: Empirical, Theoretical and Computational Perspectives* (Studies in Generative Grammar 85). Berlin: Mouton de Gruyter. https://doi.org/10.1515/9783110197549

Kertész, András and Rákosi, Csilla (2008) Introduction: The problem of data and evidence in theoretical linguistics. In András Kertész and Csilla Rákosi (eds.) *New Approaches to Linguistic Evidence: Pilot Studies / Neue Ansätze zu Linguistischer Evidenz: Pilotstudien* 9–19. Frankfurt am Main: Lang.

Kertész, András and Rákosi, Csilla (2012) *Data and Evidence in Linguistics: A Plausible Argumentation Model.* Cambridge: Cambridge University Press. https://doi.org/10.1017/CBO9780511920752

Kertész, András and Rákosi, Csilla (eds.) (2014a) *The Evidential Basis of Linguistic Argumentation* (Studies in Language Companion Series 153). Amsterdam: John Benjamins. https://doi.org/10.1075/slcs.153

Kertész, András and Rákosi, Csilla (2014b) The p-model of data and evidence in linguistics. In András Kertész and Csilla Rákosi (eds.) *The Evidential Basis of Linguistic Argumentation* (Studies in Language Companion Series 153) 15–48. Amsterdam: John Benjamins.

Kiefer, Ferenc (2006) *Aspektus és akcióminőség: Különös tekintettel a magyar nyelvre* [Aspect and Aktionsart with special regard to Hungarian]. Budapest: Akadémiai Kiadó.

Ladányi, Mária (1998) Jelentésváltozás és grammatikalizáció – kognitív és szerves nyelvelméleti keretben [Semantic change and grammaticalization in a cognitive and organic theoretical linguistic context]. *Magyar nyelv* 94(4): 407–423.

Lafont, Robert (1963–1968) Reflexions sobre lo perfach perifrastic amb *anar* en catalan e en occitan. *Estudis Romànics* 12: 271–277.

Lakoff, George and Johnson, Mark (1986) *Metáforas de la vida cotidiana*. Madrid: Cátedra.

Lass, Roger (1997) *Historical Linguistics and Language Change* (Cambridge Studies in Linguistics 81). Cambridge: Cambridge University Press. https://doi.org/10.1017/CBO9780511620928

Lehmann, Christian (2004) Data in linguistics. *The Linguistic Review* 21(3–4): 175–210. https://doi.org/10.1515/tlir.2004.21.3-4.175

Lehmann, Christian (2005) Theory and method in grammaticalization. *Zeitschrift für Germanistische Linguistik* 32(2): 152–187.

Levinson, Stephen C. (1983) *Pragmatics*. Cambridge: Cambridge University Press. https://doi.org/10.1017/CBO9780511813313

Levinson, Stephen C. (2000) *Presumptive Meanings: The Theory of Generalized Conversational Implicature*. Cambridge, MA: MIT Press. https://doi.org/10.7551/mitpress/5526.001.0001

Marquèze-Pouey, Louis (1955) L'auxiliaire *aller* dans l'expression du passé en gascon. *Via Domita* 2: 111–121.

Martínez Gómez, Esther (2004) Las perífrasis verbales en español. *Revista electrónica de estudios filológicos* 7. http://www.um.es/tonosdigital/znum7/estudios/kdelasperifrasis.htm.

Martos, Josep Lluís (1996) Cognició, temporalitat i lingüística comparada: el pretèrit perfet perifràstic i la metàfora del temps en català. *Caplletra* 20: 111–127.

Matte Bon, Francisco (1998) *Gramática comunicativa del español* (Vol. 1). Madrid: Edelsa.

Meillet, Antoine (1912/1921) L'évolution des formes grammaticales. In *Linguistique historique et linguistique générale* 130–148. Paris: Champion.

Mendeloff, Henry (1968) The Catalan periphrastic perfect reconsidered. *Romanistisches Jahrbuch* 19: 319–326. https://doi.org/10.1515/9783110244762.319

Meyer-Lübke, Wilhelm (1890–1902) *Grammatik der romanischen Sprachen*. Leipzig: Fues's Verlag.

Mitchell, Terence F. and al-Hassan, Shahir A. (1994) *Modality, Mood and Aspect in Spoken Arabic with Special Reference to Egypt and the Levant* (Library of Arabic Linguistics). London: Kegan Paul.

Moll, Francesc de B. (1991/2006) *Gramàtica històrica catalana* (Biblioteca Lingüística Catalana 31). València: Universitat de València.

Nadal, Josep M. and Prats, Modest (1982/1996) *Història de la llengua catalana*. Barcelona: Edicions 62.

Nagy C., Katalin (2008a) Aspecto verbal en la evolución de la construcción catalana '*anar* + infinitiu'. In Tibor Berta (ed.) *Acta Hispánica 13* 75–96. Szeged: Hungaria.

Nagy C., Katalin (2008b) Data in historical pragmatics: A case study on the Catalan periphrastic perfective past. In András Kertész and Csilla Rákosi (eds.) *New Approaches to Linguistic Evidence: Pilot Studies* (MetaLinguistica 22) 171–197. Frankfurt am Main: Lang.

Nagy C., Katalin (2009) Pragmatic and cognitive aspects of the research into grammaticalization. *Argumentum* 5: 110–127.

Nagy C., Katalin (2010a) The pragmatics of grammaticalisation: The role of implicatures in semantic change. *Journal of Historical Pragmatics* 11(1): 67–95. https://doi.org/10.1075/jhp.11.1.03nag

Nagy C., Katalin (2010b) The cognitive background of grammaticalization. In Enikő Németh T. and Károly Bibok (eds.) *The Role of Data at the Semantics-Pragmatics Interface* (Mouton Series in Pragmatics 9) 207–260. Berlin: Mouton de Gruyter.

Nagy C., Katalin (2014) Methods and argumentation in historical linguistics: A case study. In András Kertész and Csilla Rákosi (eds.) *The Evidential Basis of Linguistic Argumentation* (Studies in Language Companion Series 153) 71–102. Amsterdam: John Benjamins.

Narrog, Heiko and Heine, Bernd (eds.) (2011) *The Oxford Handbook of Grammaticalization*. Oxford: Oxford University Press.

Navarro, Federico (2008) Análisis histórico del discurso: Hacia un enfoque histórico-discursivo en el estudio diacrónico de la lengua. In Antonio Moreno Sandoval (ed.) *El valor de la diversidad [meta]lingüística: Actas del VIII Congreso de Lingüística General*. Madrid: Universidad Autónoma de Madrid. http://www.lllf.uam.es/clg8/actas/pdf/paperCLG85.pdf.

Nemesi, Attila László (2006) Szó szerinti jelentés, konvencionális jelentés, vezérjelentés [Literal meaning, conventional meaning, and the most salient meaning]. *Világosság* 8–10: 31–43.

Németh T., Enikő (1994) Megnyilatkozás: Típus–példány [Utterance: Type and token]. *Néprajz és Nyelvtudomány* 35: 69–101.

Németh T., Enikő (2006) Book review of Ira A. Noveck and Dan Sperber (eds.) 2004: Experimental Pragmatics. Houndmills, Basingstoke/New York, Palgrave Macmillan. *Intercultural Pragmatics* 3(1): 117–127.

Németh T., Enikő (2015) The role of perspectives in various forms of language use. *Semiotica* 203: 53–78. https://doi.org/10.1515/sem-2014-0072

Neves, Maria Helena de Moura (2000) *Gramática de usos do português*. São Paulo: Unesp.

Olbertz, Hella (1998) *Verbal Periphrases in a Functional Grammar of Spanish* (Functional Grammar Series 22). Berlin: Mouton de Gruyter. https://doi.org/10.1515/9783110820881

Orozco, Rafael (2007) Social constraints on the expression of futurity in spanish-speaking urban communities. In Jonathan Holmquist, Augusto Sayahi and Lotfi Sayahi (eds.) *Selected Proceedings of the Third Workshop on Spanish Linguistics* 103–112. Somerville, MA: Cascadilla Proceedings Project.

Penke, Martina and Rosenbach, Anette (eds.) (2004/2007) *What Counts as Evidence in Linguistics: The Case of Innateness* (Benjamins Current Topics 7). Amsterdam: John Benjamins.

Perea, Maria Pilar (2007) El perfet simple: Extensió i ús a començament del segle XXI. In Martí Sadurní (ed.) *Actes del tretzè col·loqui internacional de llengua i literatura catalanes* (Vol. 2) 279–296. Barcelona: Publicacions de l'Abadia de Montserrat.

Pérez Saldanya, Manuel (1996) Gramaticalització i reanàlisi: El cas del perfet perifràstic en català. In Axel Schönberger and Tilbert Dídac Stegmann (eds.) *Actes del desè col·loqui internacional de llengua i literatura catalanes* (Vol. 3) 71–107. Barcelona: Publicacions de l'Abadia de Montserrat.

Pérez Saldanya, Manuel (1998) *Del llatí al català: Morfosintaxi verbal històrica*. València: Universitat de València.

Pérez Saldanya, Manuel and Hualde, José Ignacio (2003) On the origin and evolution of the Catalan periphrastic preterit. In Claus D. Pusch and Andreas Wesch (eds.) *Verbalperiphrasen in den (ibero-)romanischen Sprachen* 47–60. Hamburg: Helmut Buske.

Pinkster, Harm (1990) Is the Latin present tense the unmarked, neutral tense in the system? In Rodie Risselada (ed.) *Latin in Use: Amsterdam Studies in the Pragmatics of Latin* 63–83. Amsterdam: Gieben.

Pons Bordería, Salvador and Ruiz Gurillo, Leonor (2001) Los orígenes del conector *de todas manera*: fijación formal y pragmáticas. *Revista de Filología Española* 81(3/4): 317–351. https://doi.org/10.3989/rfe.2001.v81.i3/4.180

Radatz, Hans-Ingo (2003) Las perífrasis <VADO + infinitivo> en castellano, francés y catalán: Por la misma senda – pero a paso distinto. In Claus D. Pusch and Andreas Wesch (eds.) *Verbalperiphrasen in den (ibero)romanischen Sprachen* 61–75. Hamburg: Helmut Buske.

Rescher, Nicholas (1976) *Plausible Reasoning*. Assen: Van Gorcum.

Robbeets, Martine Irma and Cuyckens, Hubert (eds.) (2013a) *Shared Grammaticalization: With Special Focus on the Transeurasian Languages*. Amsterdam: John Benjamins. https://doi.org/10.1075/slcs.132

Robbeets, Martine Irma and Hubert Cuyckens (2013b) Towards a typology of shared grammaticalization. In Martine Irma Robbeets and Hubert Cuyckens (eds.) *Shared Grammaticalization: With Special Focus on the Transeurasian Languages* 1–20. Amsterdam: John Benjamins.

Sankoff, David (1988) Sociolinguistics and syntactic variation. In Frederick Newmeyer (ed.) *Linguistics: The Cambridge Survey* (Vol. 4) 140–161. Cambridge: Cambridge University Press. https://doi.org/10.1017/CBO9780511620577.009

Scala, Luca (2003) *Català de l'Alguer: Criteris de llengua escrita*. Barcelona: Publicacions de l'Abadia de Montserrat.

Schiffrin, Deborah (1981) Tense variation in narrative. *Language* 57(1): 45–62. https://doi.org/10.1353/lan.1981.0011

Sebba, Mark (1987) *The Syntax of Serial Verbs: An Investigation into Serialisation in Sranan and Other Languages* (Creole Language Library 2). Amsterdam: John Benjamins. https://doi.org/10.1075/cll.2

Segura-Llopes, Carles (2012) El passat perifràstic en català antic. Una revisió a partir d'estudi de corpus. *eHumanista/IVITRA* 2: 118–147.

Silva-Corvalán, Carmen (1983) Tense and aspect in oral Spanish narrative: Context and meaning. *Language* 59(4): 760–780. https://doi.org/10.2307/413372

Soldevila, Ferran (1963–1968) L'ús del pretèrit perifràstic en la Crònica de Muntaner. In *Estudis de Lingüística y de Filologia Catalanes dedicats a la memòria de Pompeu Fabra en el centenari de la seva naixença* (Vol. 1) 267–270. Barcelona: Institut d'Estudis Catalans.

Sweetser, Eve E. (1988) Grammaticalisation and Semantic Bleaching. In Shelley Axmaker, Annie Jaisser and Helen Singmaster (eds.) *Proceedings of the 14th Annual Meeting of the BLS* 389–405. Berkeley, CA: Berkeley Linguistics Society.

Szertics, Joseph (1981) *Tiempo y verbo en el Romancero Viejo*. Madrid: Gredos.

Tabor, Whitney and Traugott, Elizabeth C. (1998) Structural scope expansion and grammaticalization. In Anna Giacalone Ramat and Paul J. Hopper (eds.) *The Limits of Grammaticalization* (Typological Studies in Language 37) 229–272. Amsterdam: John Benjamins. https://doi.org/10.1075/tsl.37.11tab

Tátrai, Szilárd (2010) Áttekintés a deixisről (funkcionális kognitív kiindulópontból) [A brief survey of the issue of deixis] *Magyar Nyelvőr* 134(2): 211–233.

Torres Cacoullos, Rena (2011) Variation and grammaticalization. In Manuel Díaz-Campos (ed.) *The Handbook of Hispanic Sociolinguistics* 148–167. Malden, MA: Wiley-Blackwell.

Traugott, Elizabeth C. (1999) The role of pragmatics in semantic change. In Jef Verschueren (ed.) *Pragmatics in 1998: Selected Papers from the 6th International Pragmatics Conference* (Vol. 2) 93–102. Antwerp: International Pragmatics Association.

Traugott, Elizabeth C. (2003) Constructions in grammaticalization. In Brian D. Joseph and Richard D. Janda (eds.) *The Handbook of Historical Linguistics* 624–647. Oxford: Blackwell. https://doi.org/10.1002/9780470756393.ch20

Traugott, Elizabeth C. and Dasher, Richard B. (2002/2004) *Regularity in Semantic Change.* Cambridge: Cambridge University Press.

Tsohatzidis, Savas L. (1994) Speaker meaning, sentence meaning and metaphor. In Savas L. Tsohatzidis (ed.) *Foundations of Speech Act Theory: Philosophical and Linguistic Perspectives* 365–373. London: Routledge.

Yllera, Alicia (1999/2000) Las perífrasis verbales de gerundio y participio. In Ignacio Bosque Muñoz and Violeta Demonte Barreto (eds.) *Gramática descriptiva de la lengua Española* (Vol. 2) 3391–3441. Madrid: Espasa Calpe.

Wachowicz, Teresa Cristina (2007) Auxiliary and aspectualizer verbs: Some syntactic and semantic distinctions. *Revista Letras* 73: 223–234. https://doi.org/10.5380/rel.v73i0.7555

Wolfson, Nessa (1979) The conversational historical present alternation. *Language* 55(1): 168–182. https://doi.org/10.2307/412521

Zamorano Mansilla, Juan Rafael (2006) *La generación de tiempo y aspecto en inglés y español: un estudio funcional contrastivo.* PhD Thesis. Madrid. http://www.ucm.es/BUCM/tesis/fll/ucm-t29560.pdf

Subject index

CPSIA information can be obtained
at www.ICGtesting.com
Printed in the USA
JSHW020812180220
4290JS00001B/9

9 781781 797464